NEUROMETHODS ☐ 3

Amino Acids

NEUROMETHODS
Program Editors: Alan A. Boulton and Glen B. Baker

Series I: Neurochemistry

1. **General Techniques**
   Edited by *Alan A. Boulton* and *Glen B. Baker,* 1985
2. **Amines and Their Metabolites**
   Edited by *Alan A. Boulton, Glen B. Baker,* and *Judith M. Baker,*
   1985
3. **Amino Acids**
   Edited by *Alan A. Boulton, Glen B. Baker,* and *James D. Wood,*
   1985
4. **Receptor Binding**
   Edited by *Alan A. Boulton, Glen B. Baker,* and *Pavel D. Hrdina,*
   1986
5. **Enzymes**
   Edited by *Alan A. Boulton, Glen B. Baker,* and *Peter H. Yu,* 1986

NEUROMETHODS

Series I: Neurochemistry

Program Editors: Alan A. Boulton and Glen B. Baker

NEUROMETHODS □ 3
Amino Acids

Edited by

Alan A. Boulton, Glen B. Baker,
and James D. Wood

Humana Press • Clifton, New Jersey

Library of Congress Cataloging in Publication Data

Main entry under title:

Amino acids.

  (Neuromethods; 3. Series I, Neurochemistry)
  Includes bibliographies and index.
  1. Amino acids—Analysis. 2. Neurochemistry—
Technique. I. Boulton, A. A. (Alan A.) II. Baker,
Glen B., 1947-   . III. Wood, J. D. IV. Series:
Neuromethods; 3. V. Series: Neuromethods. Series I,
Neurochemistry. [DNLM: 1. Amino acids—analysis.
2. Neurochemistry—methods. W1 NE337G v. 3 / QU 60 A516]
QP 561.A468 1985   591.19′245   85-24862
ISBN 0-89603-077-6

© 1985 The Humana Press Inc.
Crescent Manor
PO Box 2148
Clifton, NJ 07015

Printed in the United States of America

Foreword

Techniques in the neurosciences are evolving rapidly. There are currently very few volumes dedicated to the methodology employed by neuroscientists, and those that are available often seem either out of date or limited in scope. This series is about the methods most widely used by modern-day neuroscientists and is written by their colleagues who are practicing experts.

Volume 1 will be useful to all neuroscientists since it concerns those procedures used routinely across the widest range of subdisciplines. Collecting these general techniques together in a single volume strikes us not only as a service, but will no doubt prove of exceptional utilitarian value as well. Volumes 2 and 3 describe all current procedures for the analyses of amines and their metabolites and of amino acids, respectively. These collections will clearly be of value to all neuroscientists working in or contemplating research in these fields. Similar reasons exist for Volume 4 on receptor binding techniques since experimental details are provided for all types of ligand-receptor binding, including chapters on general principles, drug discovery and development, and a most useful appendix on computer programs for Scatchard, nonlinear, and competitive displacement analyses. Volume 5 provides procedures for the assessment of enzymes involved in biogenic amine synthesis and catabolism.

Volumes in the NEUROMETHODS series will be useful to neurochemists, -pharmacologists, -physiologists, -anatomists, psychopharmacologists, psychiatrists, neurologists, and chemists (organic, analytical, pharmaceutical, medicinal); in fact, everyone involved in the neurosciences, both basic and clinical.

Preface

Although it was in 1913 that Aberholden and Weil described the individual amino acid content of hydrolyzates of gray matter, white matter, spinal cord, and peripheral nerve, many years were to pass before a definitive analysis of the levels of the various free amino acids in the central nervous system was reported. The development of microbiological assays for various amino acids led in 1950 to the studies of Schurr and coworkers that provided data on the levels of 12 different amino acids in brain tissue. However, the procedures used in these investigations did not allow the determination of tissue levels of glutamate, glutamine, aspartate, glycine, alanine, or serine, among which group of amino acids are some of particular interest to neuroscientists because of their role or potential role as amino acid transmitter substances.

The microbial assays were, however, rapidly superseded by the advent of paper and ion-exchange chromatographic techniques that were based on the initial studies in the 1940s by Consden, Gordon, and Martin. Ion-exchange chromatography in particular was subject to vigorous technical improvements resulting in the use of both cationic and anionic resins for the estimation of the amino acids. By the early 1950s Partridge and coworkers in Britain, and Moore and Stein in the United States, had developed these ion-exchange chromatographic procedures to the point that the methods were adopted by a multitude of laboratories. Thus came about the "information explosion" on the amino acid levels in innumerable species and tissues, as well as on the effects of drugs and experimental conditions on these levels. The 1950s also saw the development of gas chromatography by James and Martin and of improvements in thin layer chromatography by Stahl, both methodologies eventually being incorporated into the battery of chromatographic techniques used in the determination of amino acids.

This volume deals with the major procedures currently being applied to amino acid analysis, some of them relying essentially on long-established methodologies (e.g., automated amino acid analysis; gas chromatography), others depending on comparatively recent developments based on earlier-established techniques (e.g., HPLC; gas chromatography–mass spectrometry; double-isotope dansylation), and yet others based on entirely new approaches (e.g., radioreceptor assays). In addition, this volume contains chapters on related methodologies that are vital tools for researchers working in the field of amino acids and the central nervous system. Examples of these are the techniques used to study the uptake and release of amino acids from nervous tissue, the location of the γ-aminobutyric acid (GABA) system in the central nervous system by autoradiography or by immunocytochemistry, as well as the turnover of amino acid transmitters. Some duplication may be evident between chapters. However, each author views the situation from a different perspective, and the presentation of the material from both viewpoints can be refreshing and informative.

The chapter on automated amino acid analysis describes long-established procedures, but also provides the reader with an insight into the various problems and limitations encountered with the procedures. Two chapters are devoted to the analysis of amino acids by gas chromatography procedures, one describing the use of gas chromatography *per se*, and the other describing this technique in combination with mass spectrometry. The latter combined procedure couples the excellent separative power of gas chromatography with the direct identification and quantitation capabilities of mass spectrometry. Unfortunately, the high cost of the equipment does not make it readily available to many researchers.

Another methodology described in this volume is the use of 1-dimethylaminonaphthalene-5-sulfonyl chloride (dansyl chloride) in the estimation of amino acid levels. A double-isotope dansyl microassay for cerebral amino acids is described in detail, including the use of thin-layer chromatographic techniques for the separation of the dansyl derivatives. The usefulness of the double-isotope labeling in overcoming problems inherent in the dansyl procedure is discussed by the author. Apart from a liquid scintillation counter, which is available in most laboratories, no sophisticated or expensive equipment is required for the microassay. On the other hand, extensive and time-consuming manual manipulations are required for the procedure.

The application of high performance liquid chromatography (HPLC) to the analysis of amino acids forms the subject matter of another chapter. HPLC is one of the more recent developments in amino acid determinations, and the use of precolumn fluorogenic derivatization in combination with reversed-phase liquid chromatography has simplified and improved the separation of amino acids. This system combines simplicity of sample preparation with high sensitivity and speed of analysis. Not surprisingly, HPLC is the system of choice for amino acid analysis by an ever-increasing number of laboratories.

Estimation of amino acid levels by radioreceptor binding techniques is a relatively new approach. The chapter on this topic reviews the underlying principles of ligand binding assays and details the practical considerations fundamental to the successful utilization of these techniques. Although this type of assay is extremely sensitive, it is beset with problems of specificity, and only one amino acid (GABA) can currently be determined accurately by the method. Future developments should, however, extend its applicability to some of the other amino acids, but it is likely to remain a method for determining a particular amino acid, rather than one to be utilized for the measurement of a spectrum of amino acids.

The application of immunocytochemistry to amino acid transmitter systems in the central nervous system has provided valuable information on the cellular and subcellular locations of these transmitter systems. It is particularly suited to the detection of the GABA system, and this topic forms the focal point of the chapter on immunocytochemical techniques. Although the immunocytochemical approach is extremely valuable in specific situations, the method is of limited applicability in the amino acid field, only the GABA and taurine systems having been mapped by these procedures.

A related topic, the in vitro autoradiographic localization of amino acid transmitter systems, is dealt with in a subsequent chapter. This technique involves the detection of amino acid receptors and high-affinity uptake sites using radioactively labeled ligands and light microscopy. This approach is, of course, limited to amino acids with transmitter function, and the GABA, glutamate, and glycine systems are dealt with in the chapter.

The turnover of amino acids acting as neurotransmitters has been the focus of much attention and the source of much controversy. The chapter on this subject explains how the complex compartmentation of amino acid metabolism in brain tissue gives rise to problems in interpreting "turnover" data. The results of previous studies are discussed within the framework of the known and postulated compartmentation of the amino acids.

The final chapter of this volume deals with the various techniques used to measure amino acid uptake and release both in vitro (slices and synaptosome preparations) and in vivo (cortical cups and push–pull cannulae).

In conclusion, this volume provides a detailed account of the various techniques used to determine amino acid levels, together with the attendant problems, pitfalls, advantages, and disadvantages of the methods. It also provides the reader with background information and methodological approaches used to locate amino acid neurotransmitter systems and to study the dynamics of the systems. It is intended that this volume serve as a reference source for the established investigator in the amino acid field and also as a guide to the new investigator with respect to the selection of the appropriate techniques.

Alan A. Boulton
Glen B. Baker
James D. Wood

Contributors

DEREK A. APPLEGARTH • *Biochemical Diseases Laboratory, Department of Pediatrics, British Columbia Children's Hospital, Vancouver, British Columbia, Canada*

ROGER F. BUTTERWORTH • *Laboratory of Neurochemical Clinical Research Center, Hôpital Saint-Luc, University of Montréal, Montréal, Québec, Canada*

D. L. CHENEY • *Neuroscience Research, Research Department CIBA-GEIGY Corporation, Pharmaceuticals Division, Summit, New Jersey*

RONALD T. COUTTS • *Neurochemical Research Unit, Department of Psychiatry and Faculty of Pharmacy and Pharmaceutical Sciences, University of Alberta, Edmonton, Alberta, Canada*

JOHN WILLIAM FERKANY • *Nova Pharmaceutical Corporation, Baltimore, Maryland*

FRODE FONNUM • *Norwegian Defence Research Establishment, Division for Environmental Toxicology, Kjeller, Norway*

ANDERS HAMBERGER • *Institute of Neurobiology, University of Göteborg, Göteborg, Sweden*

CHIN-TARNG LIN • *Department of Physiology, College of Medicine, The Pennsylvania State University, Hershey, Pennsylvania*

PETER LINDROTH • *Department of Analytical and Marine Chemistry, University of Göteborg, Göteborg, Sweden*

MATS SANDBERG • *Institute of Neurobiology, University of Göteborg, Göteborg, Sweden*

JOHN A. STURMAN • *Department of Pathological Biochemistry, New York State Institute for Basic Research in Developmental Disabilities, Staten Island, New York*

WOLFGANG WALZ • *Department of Pharmacology, University of Saskatchewan, Saskatoon, Saskatchewan, Canada*

PAUL L. WOOD • *Neuroscience Research, Research Department CIBA-GEIGY Corporation, Pharmaceuticals Division, Summit, New Jersey*

JANG-YEN WU • *Department of Physiology, College of Medicine, The Pennsylvania State University, Hershey, Pennsylvania*

JUPITA M. YEUNG • *Neurochemical Research Unit, Department of Psychiatry and Faculty of Pharmacy and Pharmaceutical Sciences, University of Alberta, Edmonton, Alberta, Canada*

W. SCOTT YOUNG, III • *Laboratory of Cell Biology, National Institute of Mental Health, Bethesda, Maryland*

Contents

CHAPTER 2
GAS CHROMATOGRAPHIC ANALYSIS OF AMINO ACIDS
Ronald T. Coutts and Jupita M. Yeung

CHAPTER 3
GAS CHROMATOGRAPHY—MASS FRAGMENTOGRAPHY OF
AMINO ACIDS
Paul L. Wood and D. L. Cheney

CHAPTER 4
DOUBLE-ISOTOPE DANSYL MICROASSAY FOR CEREBRAL
AMINO ACIDS
Roger F. Butterworth

CHAPTER 5
LIQUID CHROMATOGRAPHIC DETERMINATION OF AMINO
ACIDS AFTER PRECOLUMN FLUORESCENCE DERIVATIZATION
Peter Lindroth, Anders Hamberger, and Mats Sandberg

CHAPTER 6
RADIORECEPTOR ASSAYS FOR AMINO ACIDS AND
RELATED COMPOUNDS
John William Ferkany

CHAPTER 7
IMMUNOCYTOCHEMICAL TECHNIQUES
Jang-Yen Wu and Chin-Tarng Lin

CHAPTER 8
IN VITRO AUTORADIOGRAPHIC LOCALIZATION OF AMINO ACID RECEPTORS AND UPTAKE SITES
W. Scott Young, III

CHAPTER 9
DETERMINATION OF TRANSMITTER AMINO ACID TURNOVER
Frode Fonnum

CHAPTER 10
UPTAKE AND RELEASE OF AMINO ACID
NEUROTRANSMITTERS
Wolfgang Walz

Chapter 1

Automated Amino Acid Analysis

JOHN A. STURMAN AND DEREK A. APPLEGARTH

1. Introduction

Knowledge of the existence of amino acids dates back over a century in many cases, as does knowledge of their existence in proteins (*see* Vickery and Schmidt, 1931). When amino acids were discovered, their identity was established by isolating and purifying the individual compounds and obtaining elemental analyses. After the advent of paper chromatography, this technique was used with a variety of different solvents to identify elution characteristics and demonstrate the purity of isolated compounds. Amino acids were located by the use of a reagent that produced a color with the compound. The most common reagent used for locating amino acids is ninhydrin, which produces a purple color with imino acids, a pink or yellor color with amino acids, and various intermediate colors with compounds containing an amino group and a sulfonic acid, and so on. It also reacts with small peptides such as glutathione. The techniques of paper chromatography were applied to the separation of mixtures of amino acids, such as the components of a protein after hydrolysis, and then to the separation of free amino acids in physiologic fluids and tissues. It was extended by the use of two-dimensional chromatography, in which a different solvent was used in each direction. Later, electrophoresis was employed as one of the separating techniques. Even though remarkably good separations of amino acids could be obtained with this method, quantification was a problem. After locating the amino acids by reaction with ninhydrin, the spots could be cut out, eluted with alcohol or acetone, and measured in a spectrophotometer. This procedure was

1

tedious and generally only semiquantitative, but it provided the impetus to develop a better methodology.

Ion-exchange resins in columns were first employed to separate amino acids into acidic, neutral, and basic groups, although the resolution was generally not sufficient to separate individual amino acids as well as could be achieved using paper chromatography (Block, 1949; Engliss and Fiess, 1944). Better separations were obtained using columns of starch, although they had a number of drawbacks: Samples had to be desalted prior to chromatography, the columns had a relatively small capacity, and traces of carbohydrates from the starch that hindered their use for purification purposes were eluted from the columns (Moore and Stein, 1948, 1949; Stein and Moore, 1948, 1950, 1951). Sulfonated polystyrene resins, such as Dowex-50, were soon shown to be capable of separating most of the common amino acids, and this led to the development of the newer resins employed in the modern automatic amino acid analyzers (Moore and Stein, 1951; Partridge, 1949a,b; Partridge and Brimley, 1951; Partridge and Westall, 1949; Partridge et al., 1950). At this stage of the development of amino acid analysis procedures, individual fractions were collected, processed, and reacted with ninhydrin reagent, and the resulting color measured in a spectrophotometer, with the results plotted by hand. A typical analysis took several days, but enabled a complete profile of amino acids in plasma, urine, and tissues to be obtained for the first time (Stein, 1953; Stein and Moore, 1954; Tallan et al., 1954, 1958). Now that the fundamentals of separation had been achieved, rapid elution by using constant-rate pumps, continuous-flow monitoring, and automatic recording was soon utilized, and the first automatic amino acid analyzer was developed (Spackman et al., 1958). This is still the basis of the modern automatic amino acid analyzer, even though many technical advances have been made, such as the development of better resins, faster flow rates, better buffers, greater sensitivity, and automatic sample loading (e.g., Alonzo and Hirs, 1968; Benson and Patterson, 1965a,b; Felix and Terkelson, 1973; Hamilton, 1958, 1960, 1962, 1963; Hamilton et al., 1960; Moore et al., 1958; Piez and Morris, 1960; Woods and Engle, 1960). The modern systems almost invariably use lithium buffers that allow separation of glutamine (Gln) from asparagine (Asn), a separation not usually possible with sodium buffers. The ninhydrin reaction product is usually monitored at two wavelengths (440 nm and either 570 or 590 nm), and both values continuously recorded on chart paper. Even more sensitive fluorescence detection methods are now being used.

2. General Methodology

2.1. Sample Preparation

In order to measure the free amino acids in biological tissues and fluids, all proteins must first be removed. Various methods have been employed to accomplish this, such as ultrafiltration, equilibrium dialysis, and protein precipitation. Protein precipitation is considered to be the method of choice, since it is more convenient and yields more reproducible results. A variety of protein precipitants has been employed for this purpose, the most common being picric acid, perchloric acid, trichloroacetic acid, sulfosalicylic acid, ethanol, and acetone. Picric acid is a very effective protein precipitant, but has the disadvantage of requiring an extra step to remove excess picric acid by using a Dowex 2 column prior to amino acid analysis. Its use has therefore been largely discontinued. It is also a hazardous chemical to keep in the laboratory, since it is explosive when dry. Perchloric acid, trichloroacetic acid, and sulfosalicylic acid can all be used to precipitate proteins and the supernatant fluid directly used for analysis of free amino acids. For extracting tissues, the most common methods employ 10% trichloroacetic acid, usually by blending or homogenizing with 5–10 vol of acid, followed by a high-speed centrifugation (20,000g). Plasma, cerebrospinal fluid, and urine are most commonly deproteinized using 5 vol of 3% sulfosalicylic acid as protein precipitant, again followed by centrifugation. A common alternative is to add solid sulfosalicylic acid directly to the physiological fluid as follows: 30 mg/mL plasma; 5 mg/mL cerebrospinal fluid; 10 mg/mL urine (see below, for pretreatment of urine samples). The usual practice is to include an internal standard, such as norleucine or α-amino-β-guanidino propionic acid, at the precipitation stage so that it is carried through all of the procedures with the sample. γ-Aminobutyric acid (GABA) is sometimes used as an internal standard with physiological fluids, but, of course, it cannot be used with extracts of neural tissues. The organic precipitants are less effective and less convenient since they must be removed prior to amino acid analysis. They are used only in a few special situations.

The precipitated proteins can be purified by washing with hot trichloroacetic acid to remove nucleic acids and with a variety of organic solvents to remove lipids. The proteins can then be hydrolyzed, usually with 6N HCl under vacuum at 120°C for 24 h or more, and their amino acid composition determined. There are losses of some amino acids during hydrolysis—Asn, Gln, and tryptophan (Trp) are usually lost completely, and others are lost to varying degrees (*see* Light and Smith, 1963; Moore and Stein, 1963; Roach and Gehrke, 1970)—so for precise determinations a range of hydrolysis times and/or temperatures is necessary. In addition to such losses, partial racemization of cystine to mesocystine occurs, resulting in a double peak on the amino acid chromatogram (Hirs et al., 1952, 1954; Sturman et al., 1980).

2.2. Problems With Sulfhydryl Compounds

Cyst(e)ine occurs in biological tissues and fluids, both in the free sulfhydryl form, cysteine, and in the disulfide form, cystine. In the methods of sample preparation described above, cysteine is oxidized to cystine, and the proportion of the two forms in a sample cannot be determined. A method was devised to accomplish this, involving immediate addition to the sample of iodoacetate, which rapidly reacts with cysteine, converting it quantitatively to the stable *S*-carboxymethyl derivative (Brigham et al., 1960). Cystine is unaffected by iodoacetate. Both cystine and *S*-carboxymethylcysteine can be quantified by amino acid analysis. This same strategy can be used with other sulfhydryl compounds, such as homocyst(e)ine.

An additional problem encountered with sulfhydryl amino acids is that of protein binding. Both cysteine and homocysteine form covalent disulfide bonds with the cysteine residues of proteins, and will be precipitated with the proteins, resulting in erroneous determinations of both free amino acids and protein containing amino acids (Harrap et al., 1973; Kang et al., 1979; Malloy et al., 1981a,b; Smolin and Benevenga, 1982; Smolin et al., 1983). The tripeptide glutathione is also bound in the same way. To avoid this problem, the disulfide bond should be broken with a reducing agent, such as mercaptoethanol or dithiothreitol, at neutral pH, prior to protein precipitation. Thus, by a judicious combination of the above methods, the free cysteine and cystine and protein-bound cysteine content of a biological sample can be determined (*see* later notes on identification of sulfur-containing amino acids and other compounds such as the antibiotic ampicillin).

2.3. Separation Difficulties

There are some areas of the amino acid chromatogram in which several amino acids are eluted very close to each other and in which obtaining a clear separation can be a problem. Often, an improved separation of amino acids in one region of the chromatogram comes at the expense of poorer separation of amino acids in another. Some of the more common "difficult" areas of the chromatogram are discussed below.

Eluted at the very front of the chromatogram (the most acidic amino acids) are phosphoethanolamine, cysteic acid, cysteine sulfinic acid, taurine (Tau), and glyceryl phosphoethanolamine. These compounds are not well separated, especially with the very fast amino acid analysis systems. A system has been described using $0.1M$ citric acid (DeMarco et al., 1965a,b), but for better separation of this area of the chromatogram, special anion-exchange chromatographic techniques may be needed. Fortunately, in plasma one is likely to encounter only Tau and in most samples the accurate analysis of only Tau and phosphoethanolamine should be considered.

The next difficult separation is that of threonine (Thr) and serine (Ser); the separation of these amino acids can be used as a benchmark for each ion-exchange resin being used. The separation of these two peaks will become poorer as an ion-exchange resin ages. This is not usually of great importance in analyzing plasma, but it becomes crucial if one is attempting to analyze amniotic fluid samples for Ser to diagnose nonketotic hyperglycinemia (one of the diseases discussed below). The separation of these two amino acids can be improved by lowering either the pH of the buffer or the temperature at which the column is eluted. This has undesirable effects on other areas of the chromatogram, so it is usually done only when the separation of Thr and Ser is crucial.

Another difficult area in most amino acid analyzers involves cystine, methionine (Met), and cystathionine. In this area can also be found arginosuccinic acid and mixed disulfides of cyst(e)ine with homocyst(e)ine and penicillamine. By suitable alterations of pH or temperature, one can usually manage to separate a particular compound of interest in a special patient, but again, this is at the expense of problems with other areas of the chromatogram. When analyzing a physiological fluid that may contain any or all of the mixed disulfides of cyst(e)ine with other amino acids, or that may contain cystathionine or arginosuccinic acid, separations should be checked by using standards added to the physiological fluid. Arginosuccinic acid can be confused with cystathionine, and since this is also true for many paper or thin-layer chromatographic systems used to separate amino acids, one should be careful not to make rapid identification of either of these amino acids without careful verification. Allo-isoleucine can also elute in this area of the chromatogram, but this is a potential problem only in cases of branched-chain keto acid decarboxylase deficiency (maple syrup urine disease).

The final area of concern in most amino acid analyzer systems is that involving histidine (His), 1-methyl-His, 3-methyl-His, carnosine, anserine, and homocarnosine. Homocarnosine can coelute with carnosine, which may be a problem in the analysis of cerebrospinal fluid, but should not be in the analysis of plasma.

Separation of amino acids in a particular sample, especially in the difficult regions of the chromatogram discussed above, is affected by the amount of acid present in the sample. Since most samples are now prepared with trichloroacetic acid or sulfosalicylic acid, these will be used for discussion. Clearly, the ratio of sample volume to acid with either deproteinizing reagent will affect the final pH. In general, the amount of acid used should be kept to a minimum in order to minimize pH effects on the elution profile. Thus, a given sample volume in 10% trichloroacetic acid will affect the pH of the elution buffers to a greater extent than the same sample volume prepared in 3% sulfosalicylic acid. For biological materials containing relatively low concentrations of amino acids, such as plasma, urine, and cerebrospinal fluid, deproteinization with 3% sulfosalicylic acid is, therefore, the more prudent choice (again, see below for methods of pretreatment of urine samples).

3. Amino Acids of Specific Tissues and Fluids

3.1. Neural Tissue

Of all of the free amino acids found in nervous tissue, GABA has received the most attention. This amino acid is found in high concentration in the nervous system. It is not a constituent of proteins and is the major inhibitory neurotransmitter in the vertebrate nervous system (*see* Roberts et al., 1976). It is a basic amino acid, and usually no difficulties are encountered in its separation and measurement with an amino acid analyzer.

The major free amino acid in the mature mammalian nervous system is glutamic acid (Glu). This amino acid is not unique to the nervous system and it is present in virtually all tissues. It is a protein constituent and an excitatory amino acid in the vertebrate nervous system (*see* DiChiara and Gressa, 1981). It is an acidic amino acid and again presents no problems with respect to separation and quantification.

The major free amino acid in the developing mammalian nervous system is Tau. This amino acid is present in all mammalian tissues. It is not a protein constituent and its exact function is still under investigation (*see* Barbeau and Huxtable, 1978; Huxtable and Barbeau, 1976; Huxtable and Pasantes-Morales, 1982; Schaffer et al., 1981; Sturman, 1983, Sturman and Hayes, 1980). Tau is transported axonally in optic nerves, to a much greater extent in developing nerves than in mature nerves, and it has been proposed as an inhibitory neurotransmitter. Tau is an essential amino acid for cats, and if it is not supplied in their diet, they become Tau-depleted and suffer retinal and tapetal degeneration and severely impaired retinal function. Recent research suggests that primates, including man, may also have a dietary requirement for Tau. It is an acidic amino acid, and usually no difficulties are encountered in its separation and quantification by amino acid analysis, although glycerophosphoryl ethanolamine, a compound present in substantial amounts is amphibian brain, has been found to coelute with Tau in at least one analytical system (Tachiki and Baxter, 1979).

Cystathionine, the sulfur-ether intermediate in the transulfuration pathway of Met metabolism, is another amino acid of considerable interest in nervous tissue. It is found in substantial amounts only in the nervous system, and its concentration in primate brain, including human brain, is an order of magnitude greater than in brains of other mammals (*see* Gaull et al., 1975; Tallan et al., 1983). It has been suggested that cystathionine may be involved in myelination, may act as a postsynaptic inhibitor, and may be a specific neuronal marker (Griffiths and Tudball, 1976; Key and White, 1970; Tudball and Beaumone, 1979; Volpe and Laster, 1972; Werman et al., 1966; Wisniewski et al., 1985). It is a neutral amino acid, and difficulties need not be encountered in its separation and measurement with modern amino acid analyzers, although with some of the earlier models coelution with Met was sometimes a problem.

3.2. Blood, CSF, and Urine

This discussion will first address some of the issues related to these specific physiological fluids and then focus on the interpretation of high or low values of particular amino acids.

3.2.1. Blood

Plasma and serum differ in their amino acid composition (Armstrong and Stave, 1973a; De Wolfe et al., 1967; Scriver et al., 1971). Plasma is the preferred sample since clotting may take a long time and amino acid concentrations may change during this time because of either metabolic activities or lysis of blood cells

(Scriver et al., 1971). Since many amino acid analyses are performed on children to detect inborn errors of metabolism, it is sometimes necessary to decide whether one should use capillary or venous blood. The tendency of serum to yield higher and more variable amino acid concentrations than plasma (Piez and Morris, 1960) is also true for capillary blood (Applegarth, unpublished). Although most capillary amino acid values are close to those of venous plasma, there can be large variations in some amino acids such as glycine (Gly) and Thr. In general, therefore, anticoagulated venous blood samples are preferred.

Heparin is the most commonly used anticoagulant, and we have found no problems associated with its use. Perry and Hansen (1969) have reported the occurrence of ninhydrin-positive artifacts with EDTA. The two contaminants eluted with Tau and between Met and isoleucine in their system.

Because the buffy coat of blood contains leukocytes and platelets, and these have amino acid concentrations differing from those in plasma, care must be taken in handling blood not to sample the buffy coat layer. In general, we centrifuge at 2500g for 10 min and take care in removing the plasma supernatant in order to avoid sampling platelets or leukocytes. The plasma is then deproteinized immediately using sulfosalicylic acid to avoid binding of the sulfur amino acids, cysteine and homocysteine, to plasma proteins. Increased amounts of aspartic acid (Asp), Glu, or Tau should alert one to contamination of a plasma sample by cells from the buffy coat.

One other general point concerns the question of stability of samples. We always store our acidified samples at $-20°C$ or below before analysis, and under these conditions amino acids are stable for 2–3 wk. The main stability problem is that of Gln. This amino acid decomposes at temperatures above $-20°C$. One of the problems associated with the instability of Gln is that it may decompose not only to Glu, but to pyroglutamic acid, which is ninhydrin negative. Thus the inaccuracy of Gln determinations may not be readily apparent from increases in Glu values. Finally, because many amino acid analyzers are fully automated, and samples are loaded into the machine for days prior to analysis, samples may be stored in a machine under varying conditions. All investigators should verify that Gln is stable under the conditions of analysis used. In a particular system, for instance, Gln may decompose at a rate of approximately 2–3% during the time period of each analysis. It may not be possible to report Gln values unless the sample is one of the first three run through a particular machine.

3.2.2. CSF

Because there is so much less protein in this fluid than in plasma, it is usually deproteinized with 5 mg of sulfosalicylic acid/mL of fluid and prepared as plasma. Cerebrospinal fluid is not analyzed as often as plasma, but it can be of great help to analyze it to confirm cases of nonketotic hyperglycinemia (in which there is proportionately more Gly in cerebrospinal fluid than there is in plasma) (Perry et al., 1975c). In some other disorders there may be increased amounts of a metabolite characteristic of the disease in the cerebrospinal fluid, but not in the plasma (Lunde et al., 1982).

3.2.3. Urine

One problem involved in assaying urinary amino acids on an ion-exchange amino acid analyzer is that urine contains a lot of ammonia. It is usual to remove this before analysis by adding lithium hydroxide until the pH of the urine shows that it is alkaline. The urine is then stored overnight in a vacuum desiccator over concentrated sulfuric acid. After removal of ammonia by the described procedure, the sample is reconstituted with water, and sulfosalicylic acid is added. This allows for comparability of conditions used for analysis.

Since urine is a fluid of varying concentration, it is often difficult to know how much material to apply to a column. It is usually easier to err on the side of applying too much urine, because one can then still calculate answers using the 440 nm tracing. Some authors quantify amino acids by applying a fixed amount of creatinine to each column. This may be acceptable for measuring adult concentrations, but in children the amount of creatinine varies so much during the critical first 2 yr of life (critical for assessing inborn errors of metabolism), that we find this procedure to be fallacious (Applegarth et al., 1969; Applegarth and Ross, 1975).

4. Interpretations

The next part of this chapter will include a discussion of the interpretation of amino acid patterns, with particular emphasis on plasma. The rationale behind this discussion is that most amino acid analyses are undertaken in children with a view to diagnosing inborn errors of metabolism. One is then faced with the difficulty of interpreting a particular amino acid pattern to decide whether or not this pattern is normal and, if it is abnormal, to

TABLE 1
Amino Acid Concentrations in Normal Plasma[a]

Amino acid or compound	Newborns[b]	1–3 mo[c] 3–3.5 g protein/ kg/d in diet	3 mo–6 yr[d]	6–18 yr[e]
Tau	19–265	0–53	11–93	0–240
Asp	—	0–8	3–12	0–14
Hyp	18–72	—	—	0–50
Thr	65–147	64–225	40–139	74–202
Ser	62–161	76–152	93–176	71–181
Asn	—	6–33	23–79	32–62
Glu	30–103	—	11–79	7–65
Gln	243–822	—	475–746	360–740
Pro	144–329	77–324	40–332	58–324
Gly	106–254	105–222	125–318	158–302
Ala	132–455	134–416	148–475	193–545
Cit	3–36	6–36	8–47	19–52
α-Amino-*n*-butyric acid	Trace	1–34	12–43	8–36
Val	78–264	96–291	85–334	156–288
Cystine	26–71	11–38	23–68	36–58
Met	6–36	4–39	5–34	16–37
Isoleucine	27–80	32–87	13–81	38–95
Leu	61–183	43–165	40–158	79–174
Tyr	32–124	29–135	24–105	43–88
Phe	16–71	24–80	34–101	39–76
Trp	17–71	—	12–69	—
Orn	38–207	26–117	27–96	19–81
Lys	.71–272	37–168	85–218	108–233
His	32–107	42–83	22–108	64–106
Arg	17–119	21–74	32–142	44–130

[a]Amino acid concentrations are given in μmol/L.
[b]From Pohlandt, 1978.
[c]From Snyderman et al., 1968.
[d]From Applegarth et al., 1979.
[e]From Armstrong and Stave, 1973.

decide if it is sufficiently abnormal to indicate a disease state. For reasons of space, the list can not be comprehensive, but it should address the most common problems encountered in investigating potential disorders of amino acid metabolism.

We have included, for convenience, a table of normal values of plasma, cerebrospinal fluid, and urinary amino acids taken from published data (Tables 1–4). These will provide fairly

TABLE 2
Amino Acid Concentration in Normal
Cerebrospinal Fluid[a]

Amino acid or compound	Children,[b] 3 mo–10 yr	Adults[c]
Tau	3.1–16.1	1.5–11.7
Asp	—	0–1.2
Hyp	ND[d]	—
Thr	11.3–45.8	6.5–56.3
Ser	25.0–56.0	8.8–44.8
Asn	3.0–18.9	1.0–16.0
Glu	—	0–3.9
Gln	—	246–958
Pro	—	ND
Gly	1.9–10.1	1.3–11.5
Ala	12.9–53.4	10.5–49.5
Cit	—	0–4.5
α-Amino-n-butyric acid	—	0–9.7
Val	7.6–20.3	4–32.2
Cystine	Trace	0–0.7
Met	0.7–5.0	0.3–5.7
Isoleucine	1.3–7.7	0.3–9.3
Leu	3.3–21.3	1.4–23.6
Tyr	5.3–18.8	1.7–14.3
Phe	4.2–38.3	0–10.2
Trp	—	0–3.1
Orn	1.9–7.9	0.2–11.2
Lys	5.9–37.1	12.2–42.8
His	0.5–17.0	4.6–20.8
Arg	8.0–29.4	6.8–32.6

[a]Amino acid concentrations are given in μmol/L.
[b]From Applegarth et al., 1979.
[c]From Perry et al., 1975b.
[d]ND, not detectable.

specific terms of reference for comparison with data from a patient to assess clinical significance of the patient results.

In general, clinical significance depends on the amino acid level in question being at least 1.5–2 × normal. There are some exceptions to this, mostly concerning Gly and Gln, which are dealt with below. If only one or two amino acid values are high, it is usually a good idea to check one of the recognized textbooks

TABLE 3
Amino Acid Concentrations in Normal Urine[a,b]

Amino acid or compound	0–1 mo	1–6 mo	6–12 mo	1–2 yr	2–4 yr	4–7 yr	7–10 yr	10–13 yr	13 yr
Thr	218–1486	254–1379	206–1146	170–976	149–810	107–655	108–479	107–378	105–519
Ser	287–3844	382–2778	342–2117	379–1565	338–1202	225–901	183–894	141–705	125–631
Pro	70–2300	0–1095	0–300	0–270	0–220	0–120	0–60	0–60	0–60
Gly	1017–10417	1315–8804	1422–5754	1025–4596	1026–4310	761–3119	497–1713	256–2105	422–2063
Ala	554–2957	613–2874	428–2064	389–1492	255–1726	212–970	144–594	135–899	119–574
Val	38–230	63–237	61–331	40–183	59–191	41–95	26–110	21–113	22–74
Cystine	64–451	66–375	70–316	53–244	62–246	59–246	42–174	52–151	41–196
Leu	41–220	26–209	31–183	28–136	34–217	20–105	23–88	27–90	20–59
Tyr	83–401	103–690	103–944	96–542	111–591	113–480	114–342	96–383	77–239
Phe	62–220	49–391	107–367	57–314	80–306	43–260	35–159	28–146	27–120
Orn	0–210	0–275	0–125	0–125	0–125	0–105	0–105	—	0–105
Lys	454–2313	284–1507	391–1661	352–1083	279–1017	144–782	295–963	106–819	200–887
His	365–2857	727–3167	877–3346	850–3005	1009–2524	585–2250	277–1657	346–1825	168–1422
Arg	20–271	0–233	0–75	0–75	0–75	6–60	6–60	—	6–60

[a]Amino acid concentrations are given in μmol/g creatinine.
[b]From Parvy et al. (1979).

TABLE 4
Amino Acid Concentration in Normal Urine[a]

Amino acid or compound	1–7 wk[b] (n = 5)	3–12 yr[c] (n = 12)	Adult[d,e]
Tau	26–157	63–97	687–2349
Asp	—	<37	<548
Thr	13–100	85–249	126–392
Ser	59–235	155–540	257–695
Pro	28–96	—	<87
Glu	2–10	—	<68
Cit	Trace	—	0–Trace
Gly	194–787	165–1420	906–1319
Ala	46–104	102–439	236–757
α-Amino-*n*-butyric acid	Trace	16–77	—
Val	12–27	15–51	Trace–57
Cystine	9–14	21–28	42–88
Met	1–13	20–95	<67
Isoleucine	Trace–3	18–56	92–214
Leu	7–15	21–83	69–183
Tyr	22–40	40–168	83–270
Phe	7–10	24–106	55–188
β-Aminoisobutyric acid	2–7	—	—
Orn	17–21	<38	Trace–53
Lys	39–75	64–642	48–328
1-Methylhistidine	5–10	24–273	—
His	103–249	306–1285	728–2062
3-Methylhistidine	7–21	56–2485	—
Trp	—	—	50–196
Arg	Trace–7	29	<57

[a]Amino acid concentrations are given in μmol/24 h.
[b]From Levy et al., 1970.
[c]From Carver and Paska, 1971.
[d]From Tsai et al., 1980.
[e]The spread of published values is large for adult urine [*see* Shih (1973, p. 40) for a compendium of such published values].

(Stanbury et al., 1983) to see whether there is a disease associated with those amino acids. If an individual amino acid value is less than 1.5 × normal or the patient data involves many amino acid abnormalities, it is a good idea to look at patterns. For instance, a mildly increased plasma Gly along with decreased levels of Val, Leu, and/or isoleucine usually indicates starvation in the patient. Increases of Tyr and Met together may indicate hereditary tyrosinemia. An elevation of seven or eight amino acids may indicate some degree of mitochondrial dysfunction, such as that found in Reye's syndrome. We will return to such patterns later and will now concentrate on amino acids in alphabetical order.

4.1. Alanine

Ala is a metabolite of pyruvic acid, so a high Ala level should automatically lead to an assay of plasma lactate or pyruvate.

A high Ala level in combination with a high Cit suggests pyruvate carboxylase deficiency (Robinson et al., 1984).

4.2. Arginine

A common pitfall is to find a high Arg level in a urine sample taken after a patient has been given an Arg load for growth hormone stimulation. This problem is likely to be seen only in the interpretation of urinary amino acid patterns.

4.3. Asparagine

Asn and Gln together may be elevated in patients with a high blood ammonia. An increased level of Asn should lead to examination of the Gln level, and if both are elevated the plasma ammonia should be determined. Both Asn and Gln may be thought of as ammonia "sinks."

4.4. Aspartic Acid

Increased amounts of Asp should lead to a suspicion of contamination of the plasma by buffy coat (see above).

4.5. γ-Amino-n-Butyric Acid

γ-Amino-*n*-butyric acid is elevated in patients with severe liver disease, such as Reye's syndrome, and in alcoholic cirrhosis in adults.

4.6. Citrulline

Cit is seen in the disorder citrullinemia and may be increased in a variant of pyruvate carboxylase deficiency with no detectable pyruvate carboxylase protein (Robinson et al., 1984).

4.7. Cystine

A very low level of cystine in plasma may indicate that the sample was left too long before deproteinization. Many authors quote cystine values as half cystine because of a link with the past when amino acid analyzers were used to measure amino acids of proteins. Since cysteine occurs in proteins and yet dimerizes with itself to form cystine before analysis, values were quoted for half cystine. When assessing normal values, be sure to determine whether the reference is quoting half cystine or cystine levels. The cystine concentration is half of the half cystine concentration (a half cystine concentration of 10 mM is a cystine concentration of 5 mM). In the tables of normal values, it is cystine, not half cystine, that is quoted.

4.8. Glutamic Acid

See comments on aspartic acid above.

4.9. Glutamine

We have already commented on the relationship of Gln to ammonia and on the problems of decomposition of Gln in samples inadequately stored. The correlation of Gln with ammonia is not consistent, and Gln elevations may be fairly modest compared with increases in plasma ammonia. For example, a Gln of 900–1000 μM with an upper limit of normal of approximately 800 μM, can be seen with an extremely high plasma ammonia. Therefore, this is one amino acid in which any moderately increased value should be checked by measuring plasma ammonia.

4.10. Glycine

Gly levels will be increased in plasma in five major circumstances. The first is in nonketotic hyperglycinemia (Perry et al., 1975c; Stanbury et al., 1983). This is a very severe disease most likely to present in the first 2 d of life. To rule this out in a flaccid hypotonic infant presenting with a high plasma Gly, cerebrospinal fluid Gly should be immediately analyzed. In nonketotic hyperglycinemia there is a much greater increase of Gly in spinal fluid than in plasma (Perry et al., 1975c), and elevations of plasma Gly need not occur consistently. If this disease is suspected, an amino acid analysis of cerebrospinal fluid should be done immediately. Second, Gly levels may be high in diseases involving metabolism of organic acids, such as propionic or methylmalonic acid, and in all cases of acidosis with increased plasma Gly, urinary organic acids should be examined immediately. Third, in starvation there is a characteristic pattern with a high blood Gly and low branched-chain amino acids (Applegarth and Poon, 1975b). Fourth, we also see elevations of Gly in apparently normal children in whom there are no decreased levels of branched-chain amino acids. This may be a response to the caloric deprivation that an illness and a visit to the hospital entails. Usually, on reexamination the Gly level will revert to normal and no further action is indicated. Finally, some anticonvulsants such as valproic acid may cause increased levels of Gly and, since amino acid patterns are usually checked in children with seizures, this is another pitfall worth watching for.

4.11. Hydroxyproline

In a newborn it is not uncommon to find increased levels of hydroxyproline. In older children the presence of an elevated plasma hydroxyproline level usually indicates the benign disorder hydroxyprolinemia.

4.12. Isoleucine, Leucine, and Valine

Isoleucine, Leu, and Val are elevated in maple syrup urine disease and decreased in starvation (see above).

4.13. Lysine

For reasons that are imperfectly understood, we occasionally see elevations of Lys levels in very sick children in whom one can find no specific explanation. Lys concentrations are greatly elevated in patients with Reye's syndrome, although here it is part of a generalized elevation of many amino acids (Hilty et al., 1974; Romshe et al., 1981). One explanation for high Lys concentrations may be damage to mitochondria in severe illnesses. In urine patterns there are disorders of Lys reabsorption such as lysinuric protein intolerance (Rajantie et al., 1981) and cystinuria. In these diseases the urinary Lys level is elevated along with those of cystine, Arg, and Orn.

4.14. Methionine

Hypermethioninemia may result from the rare deficiency of Met adenosyltransferase (Finkelstein et al., 1975; Gaull and Tallan, 1974; Gaull et al., 1981b). This disorder is apparently benign, probably because the defect is expressed only in liver (Tallan, 1979). Persistent hypermethioninemia also occurs with normal activity of hepatic Met adenosyltransferase (Gaull et al., 1981a; Jhaveri et al., 1982). Hypermethionemia is frequently caused by ingestion of D,L-Met added to soy-based formulae that are deficient in Met. The D form of the amino acid is then seen as increased plasma and urinary Met values. Hypermethionemia also occurs in homocystinuria, and in this case both homocystine and its mixed disulfide with cystine should be seen in the amino acid patterns of plasma and urine. Met and Tyr elevations occur together in hereditary tyrosinemia and severe liver disease

(Stanbury et al., 1983). Tyrosinemia is discussed below. Hypomethioninemia may occur in a specific form of homocystinuria, accompanied by cystathioninuria and methylmalonic aciduria (Levy et al., 1970).

4.15. Ornithine

Increases may occur in a disorder called gyrate atrophy of the retina (Sipila et al., 1981), and there are less well-characterized defects involving ornithinemia (*see* Stanbury et al., 1983).

4.16. Phenylalanine

Increases of plasma Phe occur in the disorder phenylketonuria (PKU). Texts on this subject should be consulted. There may be occasional increases in plasma Tyr in the newborn infant caused by a transient failure of development of the Tyr metabolic pathway. In this case, the elevation of Tyr in plasma may cause elevations of Phe secondary to the block in Tyr metabolism. Therefore, one should always verify that the plasma Tyr of a newborn infant is normal before making the diagnosis of hyperphenylalaninemia.

4.17. Proline

Pro may also be increased in newborn children and may be elevated in kidney disease. There are at least two benign disorders of Pro metabolism (Stanbury et al., 1983).

4.18. Taurine

Increases in plasma Tau should be interpreted against other possible elevations of Asp and Glu (the most likely source of a high plasma Tau level is contamination from the buffy coat).

4.19. Tyrosine

A frequent cause of increased Tyr in plasma is liver disease, but the following comments will assume that this has been excluded and that a defect in metabolism is being considered. The discussions of Met and Phe should also be read. Tyrosinemia can be caused by at least two hereditary metabolic diseases (*see* Stanbury et al., 1983, and for a good discussion of differential diagnosis of tyrosinemia, Melancon et al., 1983).

4.20. Tryptophan

There are some problems associated with the interpretation of plasma Try levels. Try is normally bound fairly tightly to plasma proteins such as albumin, and not all methods of protein precipitation give 100% recovery of Try from plasma. Investigators have resorted to such techniques as treatment of the plasma with the detergent sodium dodecyl sulfate (SDS) prior to sulfosalicylic acid precipitation to liberate the albumin-bound Try (Mondino et al., 1975). It has been reported that deproteinization of plasma with a 10% solution of trichloroacetic acid gives close to 100% recovery (Berridge et al., 1971). Try is a puzzling amino acid, and it is interesting that in patients with analbuminemia there is a normal plasma Try level. The amounts seen in spinal fluid may be important. There are elevations of urinary Try associated with Hartnup syndrome, and this may be important in assessing children for the presence of a metabolic disease. The Hartnup trait is now usually regarded as benign.

4.21. Unusual Amino Acids

It is much easier to interpret plasma amino acid patterns than those of urine, and as a general rule urine samples need not be put through amino acid analyzers until they are first looked at by two-way paper or thin-layer chromatography. The only routine need for assay of urine amino acids by automatic amino acid analysis is to monitor patients with diseases such as cystinuria, in which the quantification of specific amino acids is important.

Occasionally one may see an unidentified peak on an amino acid chromatogram. When an unknown peak is observed in plasma, one should immediately ask for a urine sample to compare the amino acid analyzer pattern of the urine with the pattern obtained by two-way paper or thin-layer chromatography. If the unidentified peak is not seen in the urine, there might be a drug bound in plasma to a protein and this drug is released by deproteinizing the plasma. This will lead to questions about medications. Antibiotics such as penicillin or ampicillin will give unusual peaks in both blood and urine (Nyhan et al., 1972; Stegink et al., 1972). Unknowns in plasma can often be detected by an unusual 590–440 nm (or 570 to 440) ratio, because this ratio can be different from the authentic compound eluting in the same place. When this is observed, a urine sample should again be analyzed because urine will tend to concentrate many amino acids and may provide a better sample for the isolation and identification of the unknown compound.

4.22. Sulfur-Containing Amino Acids (Including Ampicillin)

The sulfur-containing compounds cystathionine, cystine, and homocystine can cause problems during amino acid analysis because they tend to elute in areas of the chromatogram that are "crowded," and they can form disulfide links with other sulfur-containing compounds. As we have already said, they can be missed because of disulfide bond formation with proteins if plasma is not deproteinized immediately. In general, therefore, when looking initially for a disease involving compounds such as homocystine and cystathionine, it is preferable to look first at a urine sample by qualitative paper, or thin-layer, two-dimensional chromatography. By using the chemical properties of the sulfur-containing amino acids, one can go a long way toward identifying each one. We usually chromatograph a urine sample on paper or thin-layer sheet before and after treatment of the urine sample with hydrogen peroxide. Although homocystine may be quite difficult to detect on a paper chromatogram, because it may not be easily separated from some other amino acids, after peroxide treatment it forms homocysteic acid, which does chromatograph in a clear and recognizable area of the chromatogram. Similarly, cystine oxidizes to cysteic acid, which can be identified. Cystathionine, which chromatographs close to cystine in many systems, is oxidized to cystathionine sulfoxide, and this is also in a clear area of the chromatogram. Met similarly forms Met sulfoxide. By a comparison of the chromatographic pattern of the urine before and after peroxide treatment, it is usually easy to see whether or not the compounds mentioned above were present in the urine. Sulfur-containing antibiotics such as ampicillin can cause ninhydrin-positive spots in a variety of places on a chromatogram (Nyhan et al., 1972; Stegink et al., 1972), but ampicillin, when treated with hydrogen peroxide, forms a compound that behaves like penicillamine sulfonic acid. This compound can be identified (Applegarth et al., 1970). Since the peroxide oxidation product of ampicillin is much more sensitive to ninhydrin than ampicillin itself, one sees a striking difference between the peroxide-treated and untreated samples. The appearance of a rather large spot in the area of penicillamine sulfonic acid is usually the clue one needs to check whether or not the patient is receiving ampicillin.

To return to the behavior of sulfur amino acids on automatic amino acid analyzers, as we have already said, cystathionine can coelute on an amino acid analyzer with arginosuccinic acid (and also with alloisoleucine). Treatment of the urine with peroxide followed by two-dimensional paper or thin-layer chromatography will confirm that the compound is indeed cystathionine.

In cases of homocystinuria, one would not only see homocystine as a separate chromatographic spot, but one usually sees cystine/homocystine mixed disulfide, which usually elutes between Leu and Tyr. The exact position of this would depend on the amino acid analyzer system used. For patients with cystinuria who are being treated with penicillamine, amino acid analyzers are commonly used to quantify the amount of free urinary cystine. A free cystine concentration of above approximately 1 mM could precipitate in urine samples and lead to cystine stone formation. When the amino acid analyzer itself is being used primarily to quantitate free cystine output, paper chromatography or thin-layer chromatography is not necessary. However, one should be aware that the mixed disulfide of cystine and penicillamine will elute somewhere in the area between cystine and Met, and the amino acid analyzer being used must be capable of separating this compound from free cystine itself.

4.23. Formiminoglutamic Acid

If the urine containing formiminoglutamic acid is subjected to amino acid chromatography on paper or thin-layer systems that use an alkaline chromatography solvent, one sees only an apparently large amount of Glu (Perry et al., 1975a). This is because formiminoglutamic acid is alkali-labile, and the use of the alkaline solvent for chromatography produces Glu. On an amino acid analyzer system, however, there would not be a large increase of Glu. There would instead be a very broad peak somewhere in the vicinity of Val (Applegarth and Poon, 1975a). This is a peak of formiminoglutamic acid that partially hydrolyzes under the conditions of chromatography. A case like this is a good example of the use of paper or thin-layer chromatography in conjunction with an amino acid analyzer system to spot that a particular urine is in need of further investigation.

4.24. Aspartylglucosamine

Aspartylglucosamine is present in urine in a storage disorder called aspartylglucosaminuria. It appears as a ninhydrin-positive, brown compound in an area of the paper or thin-layer chromatogram corresponding roughly to that of the basic amino acids such as Lys or His. On amino acid chromatography, however, it coelutes with urea (Carter and Snyder, 1981). Again, as in the case of formiminoglutamic acid, one would see an unidentified compound by paper or thin-layer chromatography that was not apparent when the amino acid analyzer was used.

5. Summary

In summary, the amino acid analyzer quantification is best reserved for situations in which direct quantification is needed. Deproteinized plasma is the preferred sample for an analysis.

A useful article that can be used to predict the effect of altering column chromatography conditions on the separation of the various amino acids is that of Long and Geiger (1969). This can be consulted if there is a need to alter chromatographic conditions to separate specific amino acids, although this might not be too useful for separation of unconventional compounds such as the mixed disulfides of sulfur amino acids.

For further details of amino acids involved in specific inborn errors of metabolism, there is a recent review by Wellner and Meister (1981). A review article such as this or the new edition of *Metabolic Basis of Inherited Diseases* (Stanbury et al., 1983) should be checked first when attempting to assess the significance of a particular amino acid pattern seen in a patient. A review article of more general significance, such as that by Applegarth et al., (1983), may also be helpful. An excellent book that contains much valuable information on analytical problems and clinically relevant material is also available (Shih, 1973), and should be consulted for further background information.

References

Alonzo N. and Hirs C. H. W. (1968) Automation of sample application in amino acid analyzers. *Anal. Biochem.* **23**, 272–288.

Applegarth D. A. and Poon S. (1975a) Determination of formiminoglutamic acid by ion-exchange chromatography. *Lancet* **I**, 1346.

Applegarth D. A. and Poon S. (1975b) Intepretation of elevated blood glycine levels in children. *Clin. Chim. Acta* **63**, 49–54.

Applegarth D. A. and Ross P. M. (1975b) The unsuitabilitiy of creatinine excretion as a basis for assessing the excretion of other metabolites in children. *Clin. Chim. Acta* **64**, 83–85.

Applegarth D. A., Edelsten A. D., Wong L. T. K., and Morrison B. J. (1979) Observed range of assay values for plasma and cerebrospinal fluid amino acid levels in infants and children aged 3 months to 10 years. *Clin. Biochem.* **12**, 173–178.

Applegarth D. A., Hardwick D. F., and Ross P. M. (1969) Creatinine excretion in children and the usefulness of creatinine equivalents in amino acid chromatography. *Clin. Chim. Acta* **22**, 131–134.

Applegarth D. A., Ingram P., Ross, P. M., and Sturrock S. (1970) Effect of ampicillin therapy on urinary amino acid patterns. *N. Engl. J. Med.* **282**, 1211.

Applegarth D. A., Toone, J. R., and McLeod P. M. (1983) Laboratory diagnosis of inborn errors of metabolism in children. *Pediat. Pathol.* **1**, 107–130.

Armstrong M. D. and Stave V. (1973a) A study of factors affecting validity of amino acid analyses. *Metabolism* **22**, 549–560.

Armstrong M. D. and Stave U. (1973b) A study of plasma free amino acid levels. II. Normal values for children and adults. *Metabolism* **22**, 561–569.

Barbeau A. and Huxtable R. J., eds. (1978) *Taurine and Neurological Disorders.* Raven, New York.

Benson J. V. and Patterson J. A. (1965a) Accelerated chromatographic analysis of amino acids commonly found in physiological fluids on a spherical resin of specific design. *Anal. Biochem.* **13**, 263–280.

Benson J. V. and Patterson J. A. (1965b) Accelerated automatic chromatographic analysis of amino acids on a spherical resin. *Anal. Chem.* **37**, 1108–1110.

Berridge B. J., Chao W. R., and Peters J. H. (1971) Column chromatographic analysis of tryptophan with the basic amino acids. *Anal. Biochem.* **41**, 256–264.

Block R. J. (1949) The Separation of Amino Acids by Ion Exchange Chromatography, in *Ion Exchange Theory and Application* (Nachod F. D., ed.), Academic, New York.

Brigham M. P., Stein W. H., and Moore S. (1960) The concentration of cysteine and cystine in human blood plasma. *J. Clin. Invest.* **39**, 1633–1638.

Carter R. J. and Snyder F. F. (1981) Measurement of aspartylglucosamine in physiological fluids with an amino acid analyser: Fused peak analysis with dual photometers. *Anal. Biochem.* **116**, 273–279.

Carver M. J. and Paska R. (1961) Ion-exchange chromatography of urinary amino acids. *Clin. Chim. Acta* **6**, 721–724.

DeMarco C., Coletta M., and Cavallini D. (1965a) Column chromatography of phosphoserine, phosphoethanolamine and S-sulfoglutathione and their identification in the presence of other amino acids. *J. Chromatogr.* **20**, 500–505.

DeMarco C., Mosti R., and Cavallini D. (1965b) Column chromatography of some sulphur-containing amino acids. *J. Chromatogr.* **18**, 492–497.

De Wolfe M. S., Baskurt S., and Cochrane W. A. (1967) Automatic amino acid analysis of blood serum and plasma. *Clin. Biochem.* **1**, 75–81.

DiChiara G. and Gressa G. L., eds. (1981) *Glutamate as a Neurotransmitter.* Raven, New York.

Engliss D. T. and Fiess H. A. (1944) Conduct of amino acids in synthetic ion exchangers. *Ind. Eng. Chem.* **36**, 604–609.

Felix A. M. and Terkelsen G. (1973) Total fluorometric amino acid analysis using fluorescamine. *Arch. Biochem. Biophys.* **157**, 177–182.

Finkelstein J. D., Kyle W. E., and Martin J. J. (1975) Abnormal methionine adenosyltransferase in hypermethioninemia. *Biochem. Biophys. Res. Commun.* **66**, 1491–1497.

Gaull G. E. and Tallan H. H. (1974) Methionine adenosyltransferase deficiency: New enzymatic defect associated with hypermethioninemia. *Science* **186**, 59–60.

Gaull G. E., Bender A. N. Vulovic D., Tallan H. H., and Schaffner F. (1981a) Methionemia and myopathy: A new disorder *Ann. Neurol.* **9**, 423–432.

Gaull G. E., Tallan H. H., Lonsdale D., Przyrembel H., Schaffner F., and von Bassewitz D. B. (1981b) Hypermethioninemia associated with methionine adenosyltransferase deficiency: Clinical morphologic, and biochemical observations on four patients. *J. Pediat.* **98**, 734–741.

Gaull G. E., Tallan H. H., Lajtha A., and Rassin D. K. (1975) Pathogenesis of Brain Dysfunction in Inborn Errors of Amino Acid Metabolism, in *Biology of Brain Dysfunction* (Gaull G. E. ed.), vol. 3, Plenum, New York.

Griffiths R. and Tudball N. (1976) Observations on the fate of cystathionine in rat brain. *Life Sci.* **19**, 1217–1223.

Hamilton P. B. (1958) Ion exchange chromatography of amino acids. Effects of resin particle size on column performance. *Anal. Chem.* **30**, 914–919.

Hamilton P. B. (1960) Ion exchange chromatography of amino acids. Study of effects of high pressures and fast flow rates. *Anal. Chem.* **32**, 1779–1781.

Hamilton P. B. (1962) Ion exchange chromatography of amino acids. Microdetermination of free amino acids in serum *Ann. N.Y. Acad. Sci.* **102**, 55–75.

Hamilton P. B. (1963) Ion exchange chromatography of amino acids. A single column, high resolving, fully automatic procedure. *Anal. Chem.* **35**, 2055–2064.

Hamilton P. B., Bogue D. C. and Anderson R. A. (1960) Ion exchange chromatography of amino acids. Analysis of diffusion (mass transfer) mechanisms. *Anal. Chem.* **32**, 1782–1792.

Harrap K. R., Jackson R. C., Riches P. G., Smith C. A., and Hill B. T. (1973) The occurrence of protein-bound mixed disulfides in rat tissues. *Biochim. Biophys. Acta* **310**, 104–110.

Hilty M. D., Romshe C. A., and Delamater P. V. (1974) Reye's syndrome and hyperaminoacidemia. *J. Pediat.* **84**, 362–365.

Hirs C. H. W., Moore S., and Stein W. H. (1952) Isolation of amino acids by chromatography on ion exchange columns; use of volatile buffers. *J. Biol. Chem.* **195**, 669–683.

Hirs C. H. W., Moore S, and Stein W. H. (1954) The amino acid composition of ribonuclease. *J. Biol. Chem.* **211**, 941–950.

Huxtable R. and Barbeau A., eds. (1976) *Taurine.* Raven, New York.

Huxtable R. J. and Pasantes-Morales H., eds. (1982) *Taurine in Nutrition and Neurology.* Plenum, New York.

Jhaveri B. M., Buist N. R. M., Gaull G. E., and Tallan, H. H. (1982) Intermittent hypermethioninemia associated with normal hepatic methionine adenosyltranferase activity: Report of a case. *J. Inherited Metab. Dis.* **5**, 101–105.

Kang S. S., Wong P. W. K., and Becker N. (1979) Protein-bound homocyst(e)ine in normal subjects and in patients with homocystinuria. *Pediat. Res.* **13**, 1141–1143.

Key B. J. and White R. P. (1970) Neuropharmacological comparison of cystathionine, cysteine, homoserine and alpha-ketobutyric acid in cats. *Neuropharmacology* **9,** 349–357.

Levy H. L., Mudd S. H., Schulman J. D., Dreyfus P. M., and Abeles R. H. (1970) A derangement of B_{12} metabolism associated with homocystinemia, cystathioninemia, hypomethioninemia and methylmalonic aciduria. *Am. J. Med.* **48,** 390–397.

Light A. and Smith E. L. (1963) Amino Acid Analysis of Peptides and Proteins, in *The Proteins* (Neurath H., ed.), vol. 1, 2nd Ed., Academic, New York.

Long C. L. and Geiger J. W. (1969) Automatic analysis of amino acids: Effect of resin cross-linking and operational variables on isolation. *Anal. Biochem.* **29,** 265–283.

Lunde H., Gjaastad O., and Gjessing L. (1982) Homocarnosinosis: Hypercarnosinuria. *J. Neurochem.* **38,** 242–245.

Malloy M. H., Rassin D. K., and Gaull G. E. (1981a) A method for the measurement of free and bound plasma cyst(e)ine. *Anal. Biochem.* **113,** 407–415.

Malloy M. H., Rassin D. K., and Gaull G. E. (1981b) Plasma cyst(e)ine in homocyst(e)inemia. *Am. J. Clin. Nutr.* **34,** 2619–2621.

Melancon S. B., Gagne R., Grenier A., Lescault A., Dallaire L., Laberge C., and Poter M. (1983) Deficiency of Fumarylacetoacetase in the Acute Form of Hereditary Tyrosinemia With Reference to Prenatal Diagnosis, in *Pediatric Liver Disease* (Fisher M. M. and Roy C. C., eds.), Plenum, New York.

Mondino A., Bongiovanni G., and Fumeros S. (1975) A new approach for obtaining total tryptophan recovery in plasma samples deproteinized with sulphosalicylic acid. *J. Chromatogr.* **104,** 297–302.

Moore S. and Stein W. H. (1948) Photometric ninhydrin method for use in the chromatography of amino acids. *J. Biol. Chem.* **176,** 367–388.

Moore S. and Stein W. H. (1949) Chromatography of amino acids on starch columns. Solvent mixtures for the fractionation of protein hydrolysates. *J. Biol. Chem.* **178,** 53–77.

Moore S. and Stein W. H. (1951) Chromatography of amino acids on sulfonated polystyrene resins. *J. Biol. Chem.* **192,** 663–681.

Moore S. and Stein W. H. (1963) Chromatographic determination of amino acids by the use of automatic recording equipment. *Meth. Enzymol.* **6,** 819–831.

Moore S., Spackman D. H., and Stein W. H. (1958) Chromatography of amino acids on sulfonated polystyrene resins. *Anal. Chem.* **30,** 1185–1190.

Nyhan W. L., Kulovich S., Hornbeck M. E., and Bachmann C. (1972) Ampicillin and amino acid chromatography. *J. Pediat.* **81,** 1215–1216.

Partridge S. M. (1949a) Displacement chromatography on synthetic ion-exchange resins. 3. Fractionation of a protein hydrolyzate. *Biochem. J.* **44,** 521–527.

Partridge S. M. (1949b) Displacement chromatography on synthetic ion-exchange resins. 4. The isolation of glucosamine and histidine from a protein hydrolysate. *Biochem. J.* **45**, 459–463.

Partridge S. M. and Brimley R. C. (1951) Displacement chromatography on synthetic ion-exchange resins. 6. Effect of temperature on the order of displacement. *Biochem. J.* **48**, 313–320.

Partridge S. M. and Westall R. G. (1949) Displacement chromatography on synthetic ion-exchange resins. 1. Separation of organic bases and amino-acids using cation-exchange resins. *Biochem. J.* **44**, 418–428.

Partridge S. M., Brimley R. C., and Pepper K. W. (1950) Displacement chromatography on synthetic ion-exchange resins. 5. Separation of the basic amino acids. *Biochem. J.* **46**, 334–340.

Parvy P., Huang Y., and Kamoun P. (1979) Age related related reference values for urinary free amino acids: A simple method of evaluation. *J. Clin. Chem. Clin. Biochem.* **17**, 205–210.

Perry T. L. and Hansen S. (1969) Technical pitfalls leading to errors in the quantitation of plasma amino acids. *Clin. Chim. Acta* **25**, 53–58.

Perry T. L., Applegarth D. A., Evans M. E., and Hansen S. (1975a) Metabolic studies of a family with massive formininoglutamic aciduria. *Pediat. Res.* **9**, 117–122.

Perry T. L., Hansen S., and Kennedy J. (1975b) Amino acids and plasma—CSF amino acid ratios in adults. *J. Neurochem.* **249**.

Perry T. L., Urquhart N., Maclean J., Evans M. E., Hansen S., Davidson A. G. F., Applegarth D. A., Macleod P. M., and Lock J. E. (1975c) Nonketotic glycine accumulation due to absence of glycine cleavage in brain. *N. Engl. J. Med.* **292**, 1269–1273.

Piez K. A. and Morris L. (1960) A modified procedure for the automatic analysis of amino acids. *Anal. Biochem.* **1**, 187–201.

Pohlandt F. (1978) Plasma amino acid concentrations in newborn infants breast fed *ad libitum. J. Pediat.* **92**, 61–616.

Rajantie J., Simell O., and Perheenupta J. (1981) Basolateral transport defect in renal tubules. *J. Clin. Invest.* **67**, 1078–1082.

Roach D., and Gehrke C. W. (1970) The hydrolysis of proteins. *J. Chromatogr.* **52**, 393–404.

Roberts E., Chase T. N., and Tower D. B. eds. (1976) *GABA in Nervous System Function.* Raven, New York.

Robinson B. H., Oei J., Sherwood W. G., Applegarth D. A., Wong L. T. K., Haworth J., Goodyer P., Casey R., and Zaleski W. A. (1984) The molecular basis for the two different clinical presentations of classical pyruvate carboxylase deficiency. *Am. J. Human Genet.* **36**, 283–294.

Romshe C. A., Hilty M. D., McClung J., Kerzner B. and Reiner C. B. (1981) Amino acid pattern in Reye's syndrome: Comparison with clinically similar entitles. *J. Pediat.* **98**, 788–790.

Schaffer S. W., Baskin S. I., and Kocsis J. J., eds. (1981) *The Effects of Taurine on Excitable Tissues.* Spectrum, New York.

Scriver C. R., Clow C. L., and Lamm P. (1971) Plasma amino acids: Screening, quantitation and interpretation. *Am. J. Clin. Nutr.* **24,** 876–890.

Shih V. E. (1973) *Laboratory Techniques for the Detection of Hereditary Metabolic Disorders.* CRC, Boca Raton, Florida.

Sipila I., Simell O., and O'Donnell J. J. (1981) Gyrate atrophy of the choroid and retina with hyperornithinemia. Characterization of mutant liver L-ornithine: 2-oxoacid aminotransferase kinetics. *J. Clin. Invest.* **67,** 1805–1807.

Smolin L. A. and Benevenga N. J. (1982) Accumulation of homocysteine in vitamin B-6 deficiency: A model for the study of cystathionine β-synthase deficiency. *J. Nutr.* **112,** 1264–1272.

Smolin L. A., Crenshaw T. D., Kurtycz D., and Benevenga N. J. (1983) Homocyst(e)ine accumulation in pigs fed diets deficient in vitamin B-6: Relationship to atherosclerosis. *J. Nutr.* **113,** 2122–2133.

Snyderman S. E., Holt Jr., L. E., Norton P. M., Rottman E., and Phanslaker S. V. (1968) The plasma aminogram. 1. Influence of the level of protein intake and a comparison of whole protein and amino acid diets. *Pediat. Res.* **2,** 131–144.

Spackman D. H., Stein W. H., and Moore S. (1958) Automatic recording apparatus for use in chromatography of amino acids. *Anal. Chem.* **30,** 1190–1206.

Stanbury J. B., Wyngaarden J. B., Fredrickson D. S., Goldstein J. L., and Brown M., eds. (1983) *Metabolic Basis of Inherited Diseases,* 5th Ed., McGraw-Hill, New York.

Stegink L. D., Boaz D. P., Von Behran P., and Mueller S. (1972) Ampicillin and amino acid chromatography. *J. Pediat.* **81,** 1214–1215.

Stein W. H. (1953) A chromatographic investigation of the amino acid constituents of normal urine. *J. Biol. Chem.* **201,** 45–58.

Stein W. H. and Moore S. (1948) Chromatography of amino acids on starch columns. Separation of phenylalanine, leucine, isoleucine, methionine, tyrosine, and valine. *J. Biol. Chem.* **176,** 337–365.

Stein W. H. and Moore S. (1950) Chromatographic determination of the amino acid composition of proteins. *Cold Spring Harbor Symposia Quant. Biol.* **14,** 179–190.

Stein W. H. and Moore S. (1951) Electrolytic desalting of amino acids. Conversion of arginine to ornithine. *J. Biol. Chem.* **190,** 103–106.

Stein W. H. and Moore S. (1954) The free amino acids of human blood plasma. *J. Biol. Chem.* **211,** 915–926.

Sturman J. A. (1983) Taurine in Nutrition Research, in *Sulfur Amino Acids: Biochemical and Clinical Aspects* (Kuriyama K., Huxtable R. J., and Iwata H., eds.), Alan R. Liss, New York.

Sturman J. A. and Hayes K. C. (1980) The biology of taurine in nutrition and development. *Adv. Nutr. Res.* **3,** 231–299.

Sturman J. A., Beratis N. G., Guarini L., and Gaull G. E. (1980) Transsulfuration by human long term lymphoid lines. Normal and cystathionase-deficient cells. *J. Biol. Chem.* **255,** 4763–4765.

Tachiki K. H. and Baxter C. F. (1979) Taurine levels in brain tissue: A need for re-evaluation. *J. Neurochem.* **33,** 1125–1129.

Tallan H. H. (1979) Methionine adenosyltransferase in man: Evidence for multiple forms. *Biochem. Med.* **21,** 129–140.

Tallan H. H., Moore S., and Stein W. H. (1954) Studies on the free amino acids and related compounds in the tissues of the cat. *J. Biol. Chem.* **211,** 927–939.

Tallan H. H., Moore S., and Stein W. H. (1958) L-Cystathionine in human brain. *J. Biol. Chem.* **230,** 707–716.

Tallan H. H., Rassin D. K., Sturman J. A., and Gaull G. E. (1983) Methionine metabolism in the brain. *Handbook of Neurochemistry* (Lajtha A., ed.) vol. 3, Plenum, New York.

Tsai M. Y., Marshall J. G., and Josephson M. W. (1980) Free amino acid analysis of untimed and 24-h urine samples compared. *Clin. Chem.* **26,** 1804–1808.

Tudball N. and Beaumone A. (1979) Studies on the neurochemical properties of cystathionine. *Biochim. Biophys. Acta* **588,** 285–293.

Vickery H. B. and Schmidt C. L. A. (1931) The history of the discovery of amino acids. *Chem. Rev.* **9,** 169–318.

Volpe J. J. and Laster L. (1972) Transsulfuration in fetal and postnatal mammalian liver and brain. *Biol. Neonate* **20,** 385–403.

Wellner D. and Meister A. (1981) A survey of inborn errors of amino acid metabolism and transport in man. *Ann. Rev. Biochem.* **50,** 911–968.

Werman R., Davidoff R. A., and Aprison M. H. (1966) The inhibitory action of cystathionine. *Life Sci.* **5,** 1431–1440.

Wisniewski K., Sturman J. A., Devine E., Rudelli R., and Wisniewski H. M. (1985) Cystathionine disappearance with neuronal loss: A possible neuronal marker. *Neuropediatrics,* **16,** 126–130.

Woods K. R. and Engle R. L. (1960) Automatic analysis of amino acids. *Ann. N.Y. Acad. Sci.* **87,** 764–774.

Chapter 2

Gas Chromatographic Analysis of Amino Acids

RONALD T. COUTTS AND JUPITA M. YEUNG

1. Introduction

Amino acids are involved in many metabolic processes and in protein synthesis. In the central nervous system, they also function as neurotransmitters or neuromodulators (Davidson, 1976; Corradetti et al., 1983; Fonnum, 1981, 1984). Numerous studies have demonstrated the excitatory effects of aspartate and glutamate (Watkins and Evans, 1981); the inhibitory effects of glycine, γ-aminobutyric acid (GABA), and taurine (Schaffer et al., 1981; Lloyd et al., 1983; Roberts, 1984); and the precursor roles of tryptophan in serotonin synthesis and of tyrosine and phenylalanine in the biosyntheses of catecholamines (Sved, 1983). It is not surprising, therefore, to see an ever-increasing interest in amino acid analysis in biological samples.

Over the last decade, a considerable number of published manuscripts (in excess of 4000) have described thin layer chromatographic (TLC), ion-exchange chromatographic (IEC), gas chromatographic (GC), and high-performance liquid chromatographic (HPLC) separations of amino acids. The popularity of GC analysis has probably reached its peak, after having gradually replaced the classical IEC procedure using the automatic amino acid analyzer. Currently, a shift of momentum toward the HPLC technique is evident.

The popularity of the automatic amino acid analyzer continues to decline dramatically in spite of its ability, in the hands of an experienced operator, to resolve the whole spectrum of amino acids in a single chromatographic run. This procedure has several disadvantages. A typical chromatographic run usually requires more than 2 h, and a relatively long equilibrium period is necessary between runs. Ninhydrin is utilized in the procedure and can present problems. This reagent is sensitive to light, atmospheric oxygen, changes in pH and temperature, and a waiting period for maturation of the reagent is required. There are also problems with column efficiency. Separations of some amino acids are incomplete, especially in the initial part of the chromatogram.

With the rapid advancement of microbore and reverse phase technology in HPLC, the escalating popularity of this analytical technique is not unexpected. Generally, in HPLC no cleanup procedure is necessary for biological samples. Some of the amino acids can be analyzed directly without prior derivatization, as is the case with tyrosine, tryptophan, and their metabolites. However, the use of fluorogenic derivatization reagents is a common practice. The amino group is the only one that requires chemical modification prior to analysis.

Precolumn derivatization methods using dansyl chloride or o-phthaldialdehyde (OPA)/2-mercaptoethanol have gained wide acceptance. These derivatives offer the advantages of having a shorter analysis time and a higher detector sensitivity when compared to the postcolumn technique of ion-exchange chromatography. However, certain disadvantages remain unresolved. In the preparation of dansyl derivatives, the necessary long reaction times and high temperatures yield multiderivatives of several amino acids. Although the OPA reaction is more selective, it is limited to primary amines (and therefore is not applicable to proline and hydroxyproline), and their derivatives are stable for only 10–15 min. Consequently, each sample has to be prepared and immediately analyzed; this prohibits automation. Both derivatives are light-, heat-, and pH-sensitive. There are also certain intrinsic technical problems associated with the HPLC method. It requires complicated solvent gradients and a longer equilibrium time between runs than the well-established GC methods, yet it only gives fair chromatographic separations.

In the GC analysis of amino acids, derivatization processes are necessary and can be a disadvantage. All reactive groups of the amino acids have to be derivatized for a satisfactory GC analysis. Esterification of the carboxylic group is acid-catalyzed, which causes the degradation of glutamine (the most abundant amino acid in the CNS) and asparagine to their dicarboxylic acid counterparts (glutamate and aspartate). An alternative derivatization approach is the conversion of amino acids to oxazolidinones, cyclic amino acid derivatives. The amide groups of glutamine and asparagine survive this procedure (Husek, 1982). To date, the simultaneous GC analysis of taurine in the presence of other amino acids has not been reported.

The advantages of GC methods of analysis are low cost; accessibility of the GC instrument and relative ease of operation; availability of virtually maintenance-free chromatographic columns; the stability of the derivatives; and the sensitivity, specificity, and accuracy of the procedure. In addition, it is relatively simple to interphase GC with a mass spectrometer (MS). This permits the identification of amino acid derivatives and differentiates them from contaminants in the sample. The use of a capillary GC column with electron-capture detection (ECD) provides excellent chromatographic separations with sensitivity in the femtomole range. Simple and short chromatographic runs are routine (Blau, 1981).

2. Classification of Amino Acids

For the purpose of this review, it is necessary to classify the natural α-amino acids into two major categories: the simple amino acids that contain only the characteristic α-amino and carboxylic acid functionalities; and those amino acids that possess an additional reactive group in the side chain. It is with this latter group that problems are associated in GC analysis. The simple amino acids that are facile to analyze are glycine (Gly), alanine (Ala), valine (Val), leucine (Leu), isoleucine (Ile), phenylalanine (Phe), and methionine (Met). The secondary amino acid, proline (Pro), presents analytical problems. The amino acids with additional reactive groups include those that contain a hydroxyl group, i.e., serine (Ser), threonine (Thr), and tyrosine (Tyr); those with a second carboxyl group, i.e., aspartate (Asp), glutamate (Glu), and their amide counterparts, asparagine (Asn) and glutamine (Gln); the basic amino acids that contain an additional amino group, i.e., lysine (Lys), ornithine (Orn), arginine (Arg), tryptophan (Trp), and histidine (His); and cysteine (CysH), which contains a side-chain thiol group, and the corresponding disulfide, cystine (Cys).

In addition to these classical α-amino acids, there are some analogous compounds that are biologically important and are generally analyzed concomitantly. These include γ-aminobutyric acid (GABA), taurine (Tau), hydroxyproline (Hyp), and β-alanine (β-Ala).

3. Clinical and Neurochemical Applications

The role of amino acids as neurotransmitters or neuromodulators has been studied extensively (Fonnum, 1984; Olsen et al., 1981; Enna, 1981; Watkins and Evans, 1981; Chiara and Gessa, 1980; Bachelard, 1981), and their precursor controls of the mono-aminergic functions reviewed (Sved, 1983; Wurtman et al., 1981; Growdon, 1979). However, an extensive literature search for studies that correlate amino acid levels with neuropsychiatric disorders has revealed that only a few amino acids have been of interest. Usually, levels of only one or a few amino acids were determined in patients with a particular disorder, apparently without considering the metabolism, utilization, or interconversion dynamics of these amino acids. For example, a decreased plasma level of Trp, relative to the quantities of five neutral amino acids, was correlated with severity of depression (DeMyer et al., 1981); reduced GABA levels were found in schizophrenia (Perry et al., 1979); increased Gln levels in cerebrospinal fluid (CSF) were found to be proportional to the severity of acute psychosis in schizophrenia (Alfredsson and Sedvall, 1983); and low CSF GABA levels were observed in seizure patients (Wood et al., 1979). A claim that low CSF Glu levels are associated with schizophrenia (Kim et al., 1980) could not be confirmed (Perry, 1982b).

4. Sampling Techniques

Physiological and laboratory factors affecting the reliability of amino acid analysis in biological fluids (Van Steirteghem and Young, 1978; Manyam and Hare, 1983) and cerebral pools (Perry, 1982a) have been reviewed recently.

    The concentration of GABA has been shown to rise appreciably in rat brain (Holdiness, 1983), within 2 min, and in human brain (Perry, 1982a), within 30 min of death, unless the brain is either instantly frozen (Alderman and Shellenberger, 1974), fixed or focused microwave irradiation (Guidotti et al., 1974; Lindgren, 1983), pretreated with 3-mercaptopropionic acid (van der Heyden and Korf, 1978), or obtained by a brain-blowing technique (Veech et al., 1973). For a simple and meaningful analysis of cerebral amino acids and related compounds, the brain should be quickly removed from the skull and frozen immediately in liquid nitrogen or isopentane in dry ice. The technique of partial thawing to a waxy consistency for dissection should be strictly followed.

However, even if the specimen is instantly frozen when removed, biochemical changes may have already occurred during surgery and brain death may introduce artifacts. To minimize these problems, in vivo sampling techniques that permit local brain perfusions have been developed. These include cortical cup (MacIntosh and Oborin, 1953), push–pull cannula (Myers, 1974), chemitrode (Delgado, 1966), and dialytrode (Delgado et al., 1984). To date, these techniques have been applied only to laboratory animals.

5. Sample Selection and Preparation

Amino acid contents have been studied in a variety of biological samples. Of these, plasma, urine, cerebrospinal fluid (CSF), and brain tissues, which are accessible and clinically significant, have received most attention. Other samples investigated include amniotic fluid, blood cells, feces, sweat, saliva, hair, fingernails, and fingerprints.

Even small amounts of protein in biological samples interfere in chromatographic analysis and have to be removed. Several procedures to remove protein have been described, including equilibrium dialysis (DeMyer et al., 1981), ultrafiltration (Moller et al., 1982), centrifugation (Hagenfeldt et al., 1984), and protein precipitation. Protein precipitation is considered to be the method of choice. Various precipitants are used, the most common of which are picric acid (Frank et al., 1981), sulfosalicylic acid (Perry, 1983), trichloroacetic acid (Freeman et al., 1980), and perchloric acid (Yamamoto et al., 1982). With all these reagents, subsequent elimination of the deproteinizing agent from the sample is essential. Other milder reagents, including ethanol (Haseyawa et al., 1981), acetone (Soley and Alemany, 1980), acetonitrile (Jones and Gilligan, 1983), and hydrochloric acid (Robitaille et al., 1982), are also employed as deproteinizing reagents.

Certain precautions must be exercised if a small sample size is to be analyzed. Extraneous amino acid contaminations may come from laboratory supplies (e.g., glassware, water, butanol, methylene chloride, hydrochloric acid, ion-exchange column, and so on) or the human body (e.g., fingerprints, skin fragments, hair, dandruff, saliva) and from dust (Husek and Macek, 1975; Lee and Drescher, 1978). Peroxides present in solvents and reagents can have an adverse effect on the success and accuracy of an assay (Riddick and Burger, 1970; Drozd, 1975; Felker, 1978; Yamamoto et al., 1982). Purchasing the highest grade solvents and reagents available or purifying them prior to use is recommended (Matucha and Smolkova, 1979).

6. Isolation of Amino Acids from Biological Samples

In most cases, biological samples are deproteinized and an aliquot of the supernatant is brought to dryness. The residue can subsequently be derivatized and analyzed without any isolation procedure. Alternatively, the amino acids in an aliquot of the deproteinized sample can be isolated on an ion-exchange resin column. Dowex 50W-X8 (100–200 mesh) resin is most commonly used for this purpose (Desgres et al., 1979; Chauhan and Darbre, 1982; Gabrys and Konecki, 1983).

A typical procedure involves a glass column (5 cm × 3 mm, id) with a glass-wool plug at the bottom, packed with the resin (H^+) in water up to a height of 1 cm. An aliquot of the biological sample (e.g., 100–500 μL of urine) that has been adjusted to pH 1–2 with HCl, is layered on top of the resin. All the noncationic impurities will be washed through the resin with 2 mL of deionized water. The amino acids are eluted with 2 mL of $4M$ ammonium hydroxide at a flow rate of one drop every 5–10 s. The eluate is then dried, derivatized, and analyzed.

7. Derivatization

Amino acids are strongly polar compounds with a variety of functional groups and cannot be analyzed by a GC method without prior conversion into suitable volatile derivatives. The preparation of such derivatives should be simple, rapid, reproducible, and give high product yields. Numerous derivatization procedures have been devised and extensively reviewed (Gehrke et al., 1968; Husek and Macek, 1975; Drozd, 1975; Blau, 1981). The purpose of this current review is to highlight some of the significant derivatization methods that have been reported within the last 10 yr and that are pertinent to neurochemistry, rather than to present an exhaustive review on methodologies.

The low volatility of amino acids caused by the presence of —NH_2, —OH, —SH, and —COOH groups in the molecules renders the GC analysis of free amino acids virtually impossible. The difficulties in selecting a suitable derivative are caused largely by the wide variation in structures and chemical properties of the side chain in amino acids. Problems are encountered particularly in the preparation of volatile derivatives of His (Hall and Nagy, 1979), Trp (Gamerith, 1983a), Gln and Asn (Collins and Summer, 1978), and especially in biological samples (McBride and Klingman, 1966; Blau, 1968; Kaiser et al., 1974).

The most extensively studied derivatization procedures for amino acids are: (a) esterification followed by acylation, and (b) cyclization of the α-aminocarboxylic acids to oxazolidinones, followed by acylation. Other significant procedures include: (c) alkoxycarbonylation followed by esterification, and (d) trimethylsilylation. Comments on all four procedures are given below.

Although various analytical methods have been described in the literature, only a few were developed for clinical use. For a good routine analytical procedure, several factors have to be considered: cost (both in time and money), choice of simple reagents, simplicity of the procedure, availability of laboratory equipment, possibility of automation, and chromatographic properties of derivatives (Gamerith, 1983a,b; Desgres et al., 1979; Lindqvist and Maenpaa, 1982).

7.1. Esterification/Acylation

Of the many derivatives available for GC analysis of amino acids, trifluoroacetylated *n*-butyl esters (TAB) have been used most widely. Since the pioneering work of Gehrke's group (Gehrke et al., 1968), the technique has been reported in detail. Sample preparation (Gehrke et al., 1968; Gamerith, 1983a,b), derivative formation (Drozd, 1975; Husek and Macek, 1975), separation on capillary column (Jaeger et al., 1981; Poole and Verzele, 1978), column supports and stationary phases (Perier et al., 1980; Gabrys and Konecki, 1983; Gamerith, 1983a), applications to the analysis of biological samples (Collins and Summer, 1978; Gabrys and Konecki, 1980; Joseph, 1978; Leighton et al., 1979), and use of ^{14}C-labeled amino acids (Matucha and Smolkova, 1979) have all been described. The mass spectra of the TAB derivatives have also been reported (Leimer et al., 1977; Vetter, 1980).

A typical procedure (Fig. 1) is one in which biological samples are deproteinized (with or without clean-up steps, e.g., using Dowex 50W-X8, 100–200 mesh), and then evaporated to dryness under a stream of nitrogen or on a rotavapor at 40–50°C. To the dry residue is added 3N HCl in *n*-butanol (prepared either by bubbling anhydrous hydrogen chloride into butanol or adding acetyl chloride to butanol). Esterification is performed at 100°C for 25–30 min. After cooling to room temperature, solutions are evaporated to dryness. Trifluoroacetylation is carried out by addition

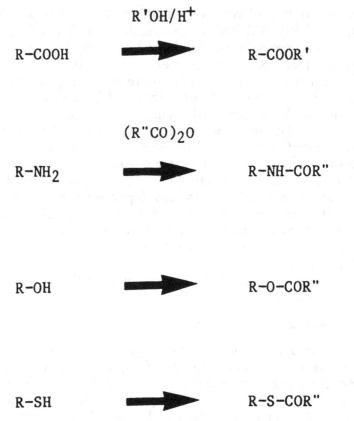

Fig. 1.   Derivatization of amino acids by esterification followed by acylation.

to the residue of trifluoroacetic anhydride in chloroform or methylene chloride and heating to 150°C for 5–10 min in a sealed tube. The mixture is then cooled to room temperature and evaporated to dryness at about 40°C. Excessive drying or evaporation at high temperature may result in loss of more volatile derivatives. The residue is then dissolved in an appropriate solvent for GC analysis using either flame-ionization (FID) or electron-capture detection (ECD).

Amino acids are also commonly esterified with 3M HC1-isobutanol heated to 120°C for 30 min (MacKenzie and Tenaschuk, 1979a), followed by acylation by reaction with heptafluorobutyric anhydride (HFBA) at 150°C for 10 min (MacKenzie and Tenaschuk, 1979b; Moodie, 1981). This procedure has been applied successfully to clinical samples (Lindqvist and Maenpaa, 1982; Chauhan et al., 1982; Desgres et al., 1979; Siezen and Mague, 1977), especially with the use of capillary columns

(Pearce, 1977; Desgres et al., 1979; Chauhan et al., 1982) and ECD (Chauhan and Darbre, 1982).

A serious limitation of this powerful analytical technique is the inability to detect and quantify Gln and Asn in an amino acid mixture. In the acid-catalyzed esterification reaction, the amides (Gln and Asn) are rapidly deaminated and converted to the same derivatives as those formed by Glu and Asp, respectively. Consequently, summations of both Glu and Gln concentrations and Asp and Asn concentrations are determined by this method. It has been shown that when Gln is heated to a temperature of about 100°C in an acidified alcoholic medium, rapid conversion to glutamic acid diester takes place. Pyrrolidone carboxylic acid ester has been shown by mass spectrometry to be a cyclized intermediate (Fig. 2). This intermediate reaches its maximum concentration after heating at 100°C for 5–10 min (Collins and Summer, 1978).

To date, a few alternative procedures have been proposed to overcome these problems. They depend on milder esterification conditions to minimize the ring opening (Collins and Summer, 1978), or the use of an enantiomeric labeling technique with D-Gln as internal standard and separation of enantiomers on achiral column (Frank et al., 1981). However, this enantiomeric separation technique is unsuccessful with Asn, since the four-membered ring intermediate (Fig. 3) does not form or cannot be trapped. Other derivatization approaches have also been studied, including cyclization to form oxazolidinones (Husek and Felt,

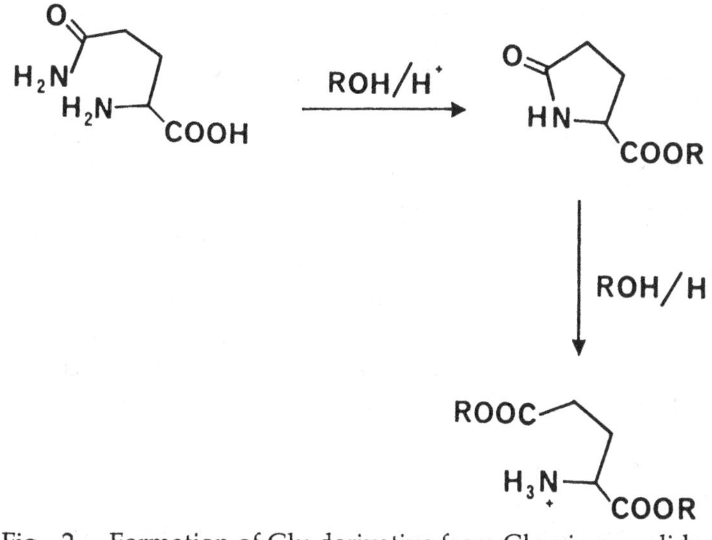

Fig. 2. Formation of Glu derivative from Gln via pyrrolidone carboxylic acid ester intermediate.

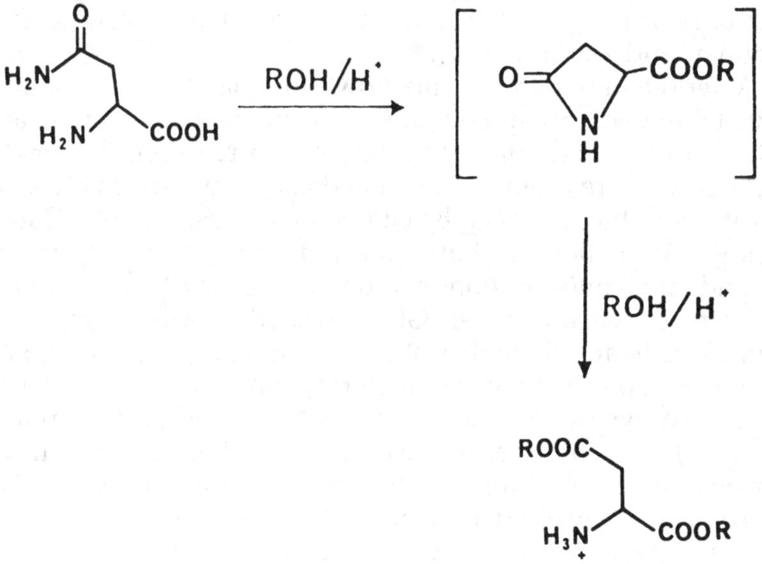

Fig. 3. Formation of Asp derivative from Asn.

1978; Husek, 1982), trimethylsilylation (Gajewski et al., 1982; Clay and Murphy, 1979), or alkoxycarbonylation (Yamamoto et al., 1982; Makita et al., 1982). Disadvantages associated with all these derivatization methods are discussed below.

7.2. Cyclization/Acylation

The use of 1,3-dichlorotetrafluoroacetone (DCTFA) in combination with heptafluorobutyric anhydride to form oxazolidinone derivatives provides a useful alternative method of analyzing amino acids (Husek and Felt, 1978, 1984; Husek, 1982; Husek et al., 1982). Since reaction conditions are mild, Gln and Asn can be analyzed in the presence of other free amino acids. The derivatization sequence involves four steps: condensation, acylation, extraction, and additional acylation (Husek, 1982). Condensation with DCTFA proceeds smoothly within 15 min at room temperature and subsequent acylation of the side chain —NH$_2$, —OH, or —SH groups is achieved by reaction with HFBA for 30 s. Critical extraction and evaporation steps are necessary for a successful assay. If quantitative analyses of Arg, Gln, and Asn are required, an additional acylation is necessary. This is achieved by heating with HFBA at 80°C for 2–3 min. The derivatization sequence is illustrated in Fig. 4 for serine.

Compared with esterification procedures, the formation of cyclic derivatives, as just described, has certain advantages: (a)

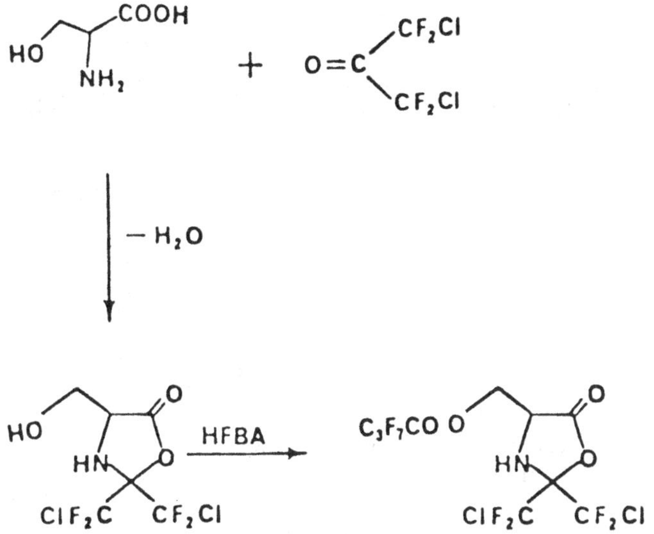

Fig. 4. Derivatization of Ser by cyclization followed by acetylation.

the cyclization reaction proceeds readily to completion at room temperature, and both the —COOH and the —NH₂ groups of α-amino acids are modified in a single chemical reaction; (b) Gln and Asn are not hydrolyzed because of the mild reaction conditions; and (c) the cyclic derivatives are sensitive to ECD. The major disadvantages of the procedure are: (a) a second column for the determination of His, Trp, and Cys is required; (b) tedious and critical extraction procedures are necessary for a successful analysis; and (c) reaction tubes have to be silanized (Husek et al., 1982). The disadvantages generally outweigh the advantages of this procedure, and this limits its application in routine clinical analysis.

7.3. Alkoxycarbonylation/Esterification

Alkoxycarbonylation followed by esterification provides an attractive alternative derivatization method for amino acid quantitation (Yamamoto et al., 1982; Makita et al., 1982). N-Alkoxycarbonyl methyl esters are prepared by a two-step procedure involving the use of an alkyl chloroformate in aqueous medium, followed by esterification with diazomethane. (Fig. 5) Alkoxycarbonylation is performed at room temperature for 10 min, whereas esterification is complete in 3 min at room temperature.

This procedure has the advantage of being able to derivatize amino acids in aqueous conditions. The N-alkoxycarbonyl deriva-

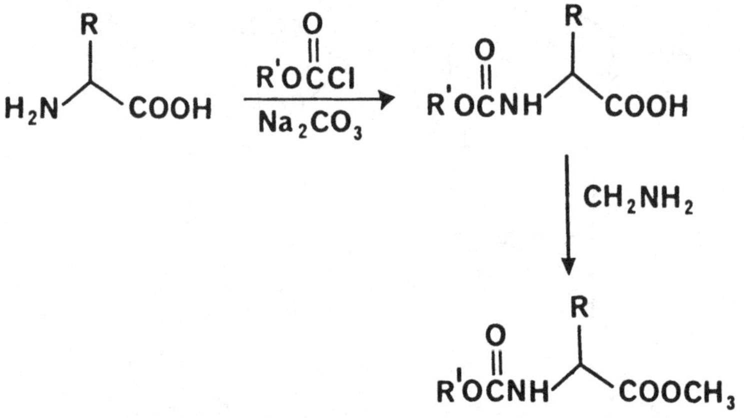

Fig. 5. Derivatization of amino acids by alkoxycarbonylation followed by esterification.

tives are extracted from the acidified reaction mixture into diethyl ether, which does not have to be dried prior to esterification. The reaction is mild and rapid; even Gln and Asn are successfully derivatized without decomposition. However, this procedure is not without disadvantages. It requires the sample to be divided into two portions, one of which is reacted with ethyl chloroformate (to derivatize Leu, Ile, Arg, and Tyr), and the other is reacted with isobutyl chloroformate (to derivatize the other amino acids). In both instances, subsequent esterification with potentially hazardous diazomethane is required. For complete separation of products, two GC columns are necessary.

7.4. Trimethylsilylation

Trimethylsilylation has been used to derivatize amino acids for GC analysis for more than two decades, but no silylation procedure of general application exists. All amino acids, including Gln and Asn, can be derivatized (Fig. 6), but problems of silylation occur with polyfunctional amino acids because of the different reactivities of —NH$_2$, —OH, —SH, and —COOH functionalities and the ready hydrolysis of the trimethylsilyl (TMS) group in the derivatives and the silylating reagents. With some of the amino acids, di- and trisubstituted TMS derivatives are formed (Husek and Macek, 1975; Drozd, 1975). In spite of these difficulties,

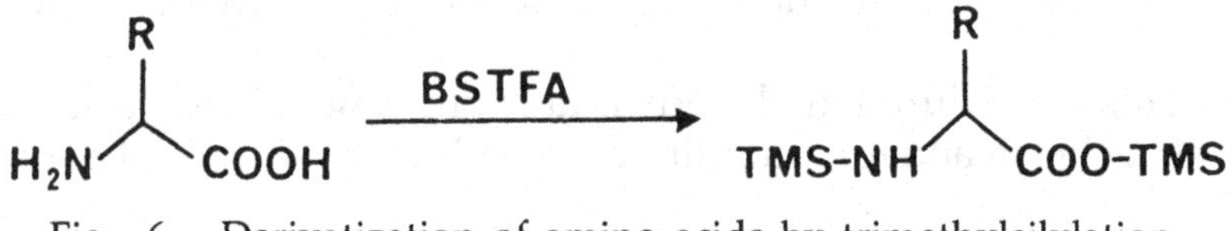

Fig. 6. Derivatization of amino acids by trimethylsilylation.

silylation procedures have been employed successfully in clinical amino acid analysis (Gajewski et al., 1982; Clay and Murphy, 1979).

Typically, the amino acids are trimethylsilylated with a mixture of bis(trimethylsilyl)trifluoroacetamide (BSTFA) and acetonitrile (1:1) by heating for 45 min at 104°C, and the TMS derivatives are analyzed on a SE–54 (12 m) GC capillary column (Gajewski et al., 1982).

7.5. Esterification/Pentafluorobenzoylation

Recently, in our laboratories, an improved method for quantitative analysis of amino acids has been developed (Yeung, Baker, and Coutts, manuscript in preparation). The carboxylic acid moiety is derivatized quantitatively by adding one drop of concentrated HCl to an isobutanol solution of the amino acids, rather than bubbling HCl gas in the conventional manner, prior to heating. Benzoyl chloride is known to be more reactive chemically than acid anhydrides, and pentafluorobenzoyl derivatives are highly sensitive to electron-capture detection (Moffat et al., 1972; Matin and Rowland, 1972; McCallum and Armstrong, 1973; Midha et al., 1979; Cristofoli et al., 1982; Nazarali et al., 1983). Pentafluorobenzoyl chloride (PFBC), therefore, was chosen to derivatize the other reactive groups of the amino acid isobutyl esters. Pentafluorobenzoylation (Fig. 7) was conducted at room temperature in aqueous conditions.

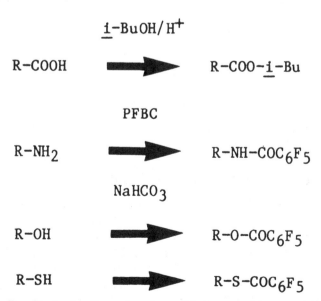

Fig. 7.   Derivatization of amino acids by esterification followed by pentafluorobenzoylation.

More than 20 amino acids can be analyzed by this procedure in less than 20 min on a fused silica capillary column with ECD. Baseline separations of all amino acid derivatives with fentomole sensitivity are routine. The excellent stability of the derivatives also means that large numbers of samples can be prepared and analyzed overnight using a GC equipped with an automatic sampler.

7.6. Analysis of Taurine

Taurine is one of the most abundant and ubiquitous free amino acids in the fluids and tissues of animals (Huxtable an Pasantes-Morales, 1982; Sturman, 1983), and much evidence suggests that it plays an important role as an inhibitory neurotransmitter or neuromodulator in the body (Davidson, 1976; Schaffer et al., 1981). The determination of Tau in biological samples has been carried out by various methods, but analysis by GC has only recently been introduced (Kataoka et al., 1984). The method involves the isocarboxycarbonylation of the amine, followed by conversion of the sulfonic acid moiety to sulfonyl chloride. The sulfonyl chloride is subsequently condensed with a secondary amine to form a sulfonamide (Fig. 8) that is analyzed on an OV-17 column. This procedure appears to be sensitive and specific for Tau.

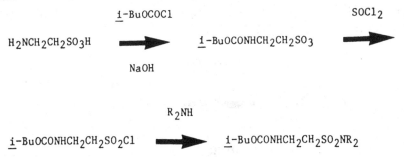

Fig. 8.   Derivatization of Tau by isobutoxycarbonylation, chlorination, and amidation.

8. Choice of Chromatographic Column

There have been dramatic changes in the GC analysis of amino acids over the last 20 yr. Earlier chromatographic separations were generally performed on packed columns, usually 1–3 m in length, with an inner diameter of 2–4 mm. With the advancement of technology, capillary columns have become more popular. The common capillary column length varies from 12–25 m, with an inner diameter of 0.2–0.5 mm. Numerous stationary phases (OV-1,

OV-11, OV-17, OV-101, QF-1, SE-54, SP-2250, EGA, Dexsil-300, and others) have been used successfully in packed columns. Extensive reviews on detectors, column supports, stationary phases, reproducibility, and separation efficiency of TAB, HFB, and TMS derivatives have been published (Husek and Macek, 1975; Blackburn, 1978) and require no further elaboration. However, the capillary columns deserve further mention. The use of various stationary phases, especially SE-30, SE-54, SE-2100, OV-1, OV-17, OV-101, OV-210, EGA, and Carbowax 20M, have been utilized; OV-101 (Chauhan et al., 1982; Chauhan and Darbre, 1982; Moodie, 1981; Husek, 1982; Desgres et al., 1979), SE—54 (Gajewski et al., 1982), and SE-30 (Poole and Verzele, 1978) are apparently superior in terms of separability, low background noise, and low column bleed.

9. Profiles of Amino Acids

Profiles of amino acids from human samples are scarce. Most investigations have employed laboratory animals and used the classical automatic amino acid analyzer. Some GC profiles of amino acids from human samples are compiled here for reference (Table 1). Laboratory animal data have been summarized elsewhere (Perry, 1982a).

10. Conclusion

Gas chromatography is a rapid, sensitive and specific method with excellent chromatographic characteristics suitable for routine clinical analysis of amino acids. When combined with mass spectrometry (GC–MS), it permits the investigator to elucidate the structures of prepared derivatives, and thus their unequivocal identification. However, the recent rapid development of high-performance liquid chromatography has provided an excellent alternative and/or complementary procedure for the analysis of amino acids.

Acknowledgments

The authors gratefully acknowledge financial support from the Alberta Heritage Foundation for Medical Research, the Medical Research Council of Canada, the Alberta Mental Health Research Fund, and Health and Welfare Canada.

TABLE 1
GC Profiles of Amino Acids in Human Samples

Amino acid	Serum,[a] mg/dL	Erythrocytes,[b] μmol/L	Urine,[c] nmol/mL	Whole blood,[d] μmol/L
Ala	3.32	331	179.8	332
Gly	2.15	362	610.2	301
Val	2.50	188	47.8	223
Pro	1.86	175	36.7	150
Asp	0.32	——[i]	215.4[f]	207[f]
Thr	1.58	124	87.1	128
Ser	1.33	164	301.4	136
Glu	0.75	——	217.9[g]	400[g]
Met	0.34	24	2.4	46
Phe	1.19	58	50.1	57[h]
Hyp	0.26	11	74.6	——
Asn	0.81	124	——	——
Gln	10.60	468	——	——
Orn	0.85	179	25.4	142
Lys	2.26	174	118.7	154
His	1.29	——	397.8	——
Trp	0.87	——	——	——
Cys	0.73	——	28.6	——
Ile	0.88	58	8.5	73
Leu	1.64	110	30.3	131
Arg	2.05	29	42.3	——
Tyr	1.12	70	63.9	103
2-ABA	——	24	60.5	——
GABA	——	12	——	——
β-Ala	——	——	5.3	——
Tau[e]				

[a]Yamamoto et al., 1982.
[b]Leighton et al., 1979.
[c]Chauhan et al., 1982.
[d]Lewis et al., 1980.
[e]499–2357 μmol/24 h urine, Kataoka et al., 1984.
[f]Asp and Asn.
[g]Glu and Gln.
[h]Phe and Hyp.
[i]— indicates valve not reported.

References

Alderman, J. L. and Shellenberger M. K. (1974) γ-Aminobutyric acid (GABA) in the rat brain: re-evaluation of sampling procedures and the postmortem increase. *J. Neurochem.* **22,** 937—940.

Alfredsson G. and Sedvall G. (1983) Rapid high-pressure liquid chromatographic method for the assay of glutamine in human cerebrospinal fluid. *J. Chromatogr.* **274,** 325–330.

Bachelard H. S. (1981) Biochemistry of Centrally Active Amino Acids, in *Amino Acid Neurotransmitters: Advances in Biochemical Psychopharmacology,* vol. 29 (Costa E. and Greengard P., eds.) Raven Press, New York, pp. 475–497.

Blackburn S. (1978) Gas Chromatographic Methods of Amino Acid Analysis, in *Amino Acid Determination,* 2nd ed. (Blackburn S., ed.) Marcel Dekker, New York, pp. 109–187.

Blau K. (1968), in *Biomedical Applications of Gas Chromatography,* vol. 2 (Szymanski H. A., ed.) Plenum Press, New York, pp. 1–52.

Blau K. (1981), in *Amino Acid Analysis* (Rattenburg J. M., ed.), Wiley, New York, pp. 48–65.

Chauhan J., Darbre A., and Carlyle R. F. (1982) Determination of urinary amino acids by means of glass capillary gas-liquid chromatography with alkali-flame ionization detection and flame ionizaton detection. *J. Chromatogr.* **227,** 305–321.

Chauhan J. and Darbre A. (1982) Determination of amino acids by means of glass capillary gas-liquid chromatography with temperature programmed electron-capture detection. *J. Chromatogr.* **236,** 151–156.

Chiara G. D. and Gessa G. L., eds. (1980) Glutamate as a Neurotransmitter, in *Advances in Biochemical Psychopharmacology,* vol. 27. Raven Press, New York.

Clay, K. L. and Murphy R. C. (1979) New procedures for isolation of amino acids based on selective hydrolysis of trimethylsilyl derivatives. *J. Chromatogr.* **164,** 417–426.

Collins F. S. and Summer G. K. (1978) Determination of glutamine and glutamic acid in biological fluids by gas chromatography. *J. Chromatogr.* **145,** 456–463.

Corradetti R., Moneti G., Moroni F., Pepeu G., and Wieraszko A. (1983) Electrical stimulation of the striatum radiatum increases the release and neosynthesis of aspartate, glutamate, and γ-aminobutyric acid in rat hippocampus slices. *J. Chromatogr.* **41,** 1518–1525.

Cristofoli W. A., Baker, G. B., Coutts R. T., and Benderly A. (1982) Analysis of a monofluorinated analogue of amphetamine in brain tissue using gas chromatography with electron-capture detection. *Progr. Neuropsychopharmacol. and Biol. Psychiat.* **6,** 373–376.

Davidson A. N. (1976) *Neurotransmitter Amino Acids.* Academic Press, London.

Delgado J. M. R. (1966) Intercerebral perfusion in awake monkeys. *Arch. Int. Pharmacodyn.* **161,** 442–462.

Delgado, J. M.. R., Lerma J., Matin del Rio R., and Solis J. M. (1984) Dialytrode technology and local profiles of amino acids in the awake cat brain. *J. Neurochem.* **42,** 1218–1228.

DeMyer M. K., Shea P. A., Hendrie H. C., and Yoshimura N. N. (1981) Plasma tryptophan and five other amino acids in depressed and normal subjects. *Arch. Gen. Psychiatry* **38,** 642–646.

Desgres J., Boisson D., and Padieu P. (1979) Gas-liquid chromatography of isobutyl ester, *N(O)*-heptafluorobutyrate derivatives of amino acids on a glass capillary column for quantitative separation in clinical biology. *J. Chromatogr.* **162,** 133–152.

Drozd J. (1975) Chemical derivatization in gas chromatography. *J. Chromatogr.* **113,** 303–356.

Enna S. J. (1981) GABA receptor pharmacology, functional considerations. *Biochem. Pharmacol.* **30,** 907–913.

Felker P. (1978) Gas-liquid chromatography of the heptafluorobutyric-*O*-isobutyl esters of amino acids. *J. Chromatogr.* **153,** 259–262.

Fonnum F. (1981) The Turnover of Transmitter Amino Acids, With Special Reference to GABA, in *Central Neurotransmitter Turnover* (Pycock C. J. and Tabener P. V., eds.) University Press, Baltimore, pp. 105–124.

Fonnum F. (1984) Glutamate: a neurotransmitter in mammalian brain. *J. Neurochem.* **42,** 1–11.

Frank H., Eimiller A., Kornhuber H. H., and Bayer E. (1981) Gas chromatographic determination of glutamine in the presence of glutamic acid in biological fluids. *J. Chromatogr.* **224,** 177–183.

Freeman M. S., Co C., Mote T. R., Lane J. D., and Smith J. E. (1980) Determination of content and specific activity of amino acids in central nervous system tissue utilizing tritium and carbon-14 dual labeling. *Anal. Biochem.* **106,** 191–194.

Gabrys J. and Konecki J. (1980) Gas-liquid chromatography of free amino acids in hyaloplasm of rat cerebral, cerebellar, and ocular tissues and in skull and heart muscle. *J. Chromatogr.* **182,** 147–154.

Gabrys J. and Konecki J. (1983) Gas-liquid chromatography of free amino acids in the cytosol of mammalian atrium and ventricle of the heart. *J. Chromatogr.* **276,** 19–27.

Gajewski E., Dizdaroglu M., and Simic M. G. (1982) Kovat's indices of trimethylated amino acids on fused-silica capillary columns. *J. Chromatogr.* **249,** 41–55.

Gamerith G. (1983a) Gas chromatography of various *N(O,S)*-acyl alkyl esters of amino acids. *J. Chromatogr.* **268,** 403–415.

Gamerith G. (1983b) Derivatization of amino acids to their *N(O,S)*-acyl alkyl esters for gas-liquid chromatographic determination. *J. Chromatogr.* **256,** 326–330.

Gehrke C. W., Roach D., Zumwalt W., Stalling D. L., and Wall L. L. (1968) *Quantitative Gas-Liquid Chromatography of Amino Acids in Proteins and Biological Substances: Macro, Semimicro, and Micro Methods.* Analytical Biochemistry Laboratories, Columbia, Missouri.

Guidotti A., Cheney D. L., Trabucchi M., Dotenchi M., and Wang C. (1974) Focused microwave radiation: a technique to minimize postmortem changes of cyclic nucleotides, dopa, and choline and to preserve brain morphology. *Neuropharmacology* **13,** 1115–1122.

Growdon J. H. (1979) Neurotransmitter Precursors in the Diet: Their Use in the Treatment of Brain Disorders, in *Nutrition and the Brain,* vol. 3 (Wurtman R. J. and Wurtman J. J., eds.) Raven Press, New York, pp. 117–181.

Hagenfeldt, L., Bjerkenstedt L., Edman G., Sedvall G., and Wiesel F.-A. (1984) Amino acids in plasma and CSF and monoamine metabolites in CSF: interrelationship in healthy subjects. *J. Neurochem.* **42,** 833–837.

Hall N. T. and Nagy S. (1979) Response amplification of histidine in gas-liquid chromatographic analysis of amino acid mixtures. *J. Chromatogr.* **171,** 392–397.

Haseyawa Y., Ono T., and Maruyama Y. (1981) γ-Aminobutyric acid in the murine brain: mass fragmentographic assay method and postmortem changes. *Japan. J. Pharmacol.* **31,** 165–173.

Holdiness M. R. (1983) Chromatographic analysis of glutamic acid decarboxylase in biological samples. *J. Chromatogr.* **277,** 1–24.

Husek P. (1982) Gas chromatography of cyclic amino acid derivatives. A useful alternative to esterification procedures. *J. Chromatogr.* **234,** 381–393.

Husek P. and Felt V. (1978) Simultaneous estimation of dicarboxylic amino acids and their amides by gas chromatography. *J. Chromatogr.* **152,** 546–550.

Husek P. and Felt. V. (1984) Possibilities and limitations in the analysis of amino acid oxazolidinones in the femtomole range by gas chromatography with electron-capture detection. *J. Chromatogr.* **305,** 442–449.

Husek P., Felt V., and Matucha M. (1982) Single column gas chromatographic analysis of amino acid oxazolidinones. *J. Chromatogr.* **252,** 217–224.

Husek P. and Macek K. (1975) Gas chromatography of amino acids. *J. Chromatogr.* **113,** 139–230.

Huxtable R. J. and Pasantes-Morales H., eds. (1982) *Taurine in Nutrition and Neurology.* Plenum Press, New York.

Jaeger H., Kloer H. U., Ditschuneit H., and Frank H. (1981) Glass capillary gas-liquid chromatography of amino acids. *Chromatogr. Sci.* **15** 331–364.

Jones B. N. and Gilligan J. P. (1983) o-Phthaldialdehyde precolumn derivatization and reversed-phase high-performance liquid chromatography of polypeptide hydrolysates and physiological fluids. *J. Chromatogr.* **226,** 471–482.

Joseph M. H. (1978) Determination of kynurenine by a simple gas-liquid chromatographic method applicable to urine, plasma, brain, and cerebrospinal fluid. *J. Chromatogr.* **146,** 33–41.

Kaiser F. E., Gehrke R. W., Zumwalt R. W., and Kuo K. (1974) *Amino Acid Analysis: Hydrolysis, Ion-Exchange Clean-Up, Derivatization, and Quantitation by Gas-Liquid Chromatography.* Analytical Biochemical Laboratories, Columbia, Missouri.

Kataoka H., Yamamoto S., and Makita M. (1984) Quantitative gas-liquid chromatography of taurine. *J. Chromatogr.* **306,** 61–68.

Kim J. S., Kornhuber H. H., Schmid-Burgk W., and Holzmuller B. (1980) Low cerebrospinal fluid glutamate in schizophrenia and a new hypothesis of schizophrenia. *Neurosci. Lett.* **20,** 379–382.

Lee S. and Drescher D. G. (1978) Fluorometric amino acid analysis with O-phthaldialdehyde (OPA). *Int. J. Biochem.* **9,** 457–467.

Lewis A. M., Waterhouse C., and Jacobs L. S. (1980) Whole blood and plasma amino acid analysis: gas-liquid and cationic exchange chromatography compared. *Clin. Chem.***26,** 271–276.

Leighton W. P., Rosenblatt S., and Chanley J. D. (1979) Determination of erythrocyte amino acids by gas chromatography. *J. Chromatogr.* **164,** 427–439.

Leimer K. R., Rice R. H., and Gehrke C. W. (1977) Complete mass spectra of *N*-trifluoroacetyl-*n*-butyl esters of amino acids. *J. Chromatogr.***141,** 121–144.

Lindgren, S. (1983) Effects of the glutamic acid decarboxylase inhibitor 3-mercaptopropionic acid on the synthesis of brain GABA in vivo and postmortally. *J. Neural Transm.* **58,** 75–82.

Lindqvist L. and Maenpaa, P. H. (1982) Gas chromatographic determination of amino acids and amino acids attached to transfer RNA in biological samples. *J. Chromatogr.* **232,** 225–235.

Lloyd K. G., DeMontis G., Broekkamp C. L., Thuret F., and Worms P. (1983) Neurochemical and Neuropharmacological Indications for the Involvements of GABA and Glycine Receptors in Neuropsychiatric Disorders, in *CNS Receptors—From Molecular Pharmacology to Behavior* (Mandel P., and DeFeudis F. V., eds.), Raven Press, new York, pp. 137–148.

MacIntosh F. C. and Oborin P. E. (1953) Release of acetylcholine from intact cerebral cortex. *Proc. XIX Int. Cong. Physiol.* pp. 580–581.

MacKenzie S. L. and Tenaschuk D. (1979a) Quantitative formation of *N(O,S)*-heptafluorobutyryl isobutyl amino acids for gas chromatographic analysis. I. Esterification. *J. Chromatogr.* **171,**195–208.

MacKenzie S. L. and Tenaschuk D. (1979b) Quantitative formation of *N(O,S)*-heptafluorbutyryl isobutyl amino acids for gas chromatographic analysis. II. Acylation. *J. Chromatogr.* **173,** 53–63.

Makita M., Yamamoto S., and Kiyama S. (1982) Improved gas-liquid chromatographic method for the determination of protein amino acids. *J. Chromatogr.* **237,** 279–284.

Manyam B. V. and Hare T. A. (1983) Cerebrospinal fluid GABA measurements: basic and clinical considerations. *Clin. Neuropharmacol.* **6,** 25–36.

Matin S. B. and Rowland M. (1972) Electron-capture sensitivity comparison of various derivatives of primary and secondary amines. *J. Pharm. Sci.* **61,** 1235–1240.

Matucha M. and Smolkova E. (1979) A new aspect of derivatization in gas chromatography of micro-amounts of amino acids. *J. Chromatogr.* **168,** 255–259.

McBride W. J. and Klingman J. D. (1966) *Lectures on Gas Chromatography* (Mattick L. R. and Szymanski H. A., eds.) Plenum Press, New York, p. 25.

McCallum N. K. and Armstrong R. J. (1973) The derivatization of phenols for gas chromatography using electron-capture detection. *J. Chromatogr.***78,** 303–307.

Midha K. K., Cooper J. K., Gagne D., and Bailey K. (1979) Detection of nanogram levels of various ring substituted phenyl-isopropylamines in urine and plasma by GLC-ECD. *J. Anal. Toxicol.* **3,** 53–58.

Moffat A. C., Horning E. C., Matin S. B., and Rowland M. (1972) Perfluorobenzene derivatives as derivatizing agents for the gas chromatography of primary and secondary amines using electron-capture detection. *J. Chromatogr.* **66,** 255–260.

Moller S. E., Kirk, L., and Honore P. (1982) Tryptophan tolerance and metabolism in endogenous depression. *Psychoneuropharmacol.* **76,** 79–83.

Moodie I. M. (1981) Gas-liquid chromatography of amino acids. The heptafluorobutyryl-isobutyl ester derivatives of tryptophan. *J. Chromatogr.* **208,** 60–66.

Myers R. D. (1974) *Handbook of Drug and Chemical Stimulation of the Brain: Behavioral, Pharmacological, and Physiological Aspects.* Van Nostrand Reinhold, New York.

Nazarali A. J., Baker G. B., Coutts R. T., and Pasutto F. M. (1983) Amphetamine in rat brain after intraperitoneal injection of N-alkylated analogues. *Progr. NeuroPsychopharmacol. and Biol. Psychiat.* **7,** 813–816.

Olsen, R. W., Bergman M. O., Van Ness P. C., Lummis S. C., Watkins A. E., Napias C., and Greenlee D. V. (1981) γ-Aminobutyric acid receptor binding in mammalian brain. Heterogeneity of binding sites. *Mol. Pharmacol.* **19,** 217–227.

Pearce R. J. (1977) Amino acid analysis by gas-liquid chromatography of N-heptafluorobutyryl isobutyl esters. Complete resolution using a support-coated tubular capillary column. *J. Chromatogr.* **136,** 113–126.

Perier C., Ronziere M. C., Rattner A., and Frey J. (1980) Employment of gas-liquid chromatography for the analysis of collagen amino acids in biopsy tissue. *J. Chromatogr.* **182,** 155–162.

Perry T. L. (1982a) Cerebral Amino Acid Pools, in *Handbook of Neurochemistry: Chemical and Cellular Architecture* (Lajtha A., ed.) Plenum Press, New York and London, pp. 151–180.

Perry T. L. (1982b) Normal cerebrospinal fluid and brain glutamate levels in schizophrenia do not support the hypothesis of glutaminergic neuronal dysfunction. *Neurosci. Lett.* **28,** 81–85.

Perry T. L. (1983) Levels of Glutamine, Glutamate, and GABA in CSF and Brain Under Pathological Conditions, in *Glutamine, Glutamate, and GABA in the Central Nervous System* (Hertz, L. Kuamme, E., McGeer, E., and Schousboe, A., eds.), Alan R. Liss, New York, pp. 581–594.

Perry T. L., Kish S. J., Buchanan J., and Hansen S. (1979) γ-Aminobutyric acid deficiency in brain of schizophrenic patients. *Lancet* **1,** 237–239.

Poole C. F. and Verzele M. (1978) Separation of protein amino acids as their N(O)-acyl alkyl ester derivatives on a glass capillary column. *J. Chromatogr.* **150,** 439–446.

Riddick J. A. and Burger W. B. (1970) *Organic Solvents,* Wiley, New York, p. 629.

Roberts E. (1984) γ-Aminobutyric Acid (GABA): From Discovery to Visualization of GABAergic Neurons in the Vertebrate Nervous System, in *Actions and Interactions of GABA and Benzodiazepines* (Bowery N. G., ed.) Raven Press, New York, pp. 1–25.

Robitaille Y., Wood P. L., Etienne P., Lal S., Finlayson M. H., Gauthier S., and Nair N. P. V. (1982) Reduced cortical choline acetyl transferase activity in Gerstmann-Straussler syndrome. *Prog. Neuropsychopharmacol. and Biol. Psychiat.* **6,** 529–531.

Schaffer S. W., Baskin S. I., and Kocsis J. J., eds. (1981) *The Effect of Taurine in Excitable Tissues.* Spectrum Publications, New York.

Siezen R. J. and Mague T. H. (1977) Gas-liquid chromatography of N-heptafluorobutyryl isobutyl esters of fifty biologically interesting amino acids. *J. Chromatogr.* **130,** 151–160.

Soley M. and Alemany M. (1980) A rapid method for the estimation of amino acid concentration in liver tissue. *J. Biochem. Biophys. Methods* **2,** 207–211.

Sturman J. A. (1983) Taurine in Nutrition Research, in *Sulfur Amino Acids: Biochemical and Clinical Aspects* (Kuriyama K., Huxtable R. J., and Iwata H., eds.) Alan R. Liss, Inc., New York, pp. 281–295.

Sved A. F. (1983) Precursor Control of the Function of Monoaminergic Neurons, in *Nutrition and the Brain,* vol. 6 (Wurtman R. J. and Wurtman J. J., eds.) Raven Press, New York pp. 223–275.

van der Heyden J. A. M. and Korf J. (1978) Regional levels of GABA in the brain: Rapid semiautomatic assay and prevention of postmortem increase by 3-mercaptopropionic acid. *J. Neurochem.* **31,** 197–203.

Van Steirteghem A. C. and Young D. S. (1978) Amino Acids in Physiological Fluids, in *Amino Acid Determination: Method and Techniques,* 2nd ed. (Blackburn S., ed.) Marcel Dekker, New York, pp. 261–317.

Veech R. L., Harris R. L. Veloso D., and Veech E. H. (1973) Freeze-blowing: a new technique for the study of brain in vivo. *J. Neurochem.* **20,** 183–188.

Vetter W. (1980) Amino acids (Mass Spectrometry), in *Biochem. Appl. Mass Spectrom.,* 1st Suppl. Vol. (Waller G. R. and Dermer O. C., eds.) Wiley, New York, pp. 439–467.

Watkins J. C. and Evans R. H. (1981) Excitatory amino acid transmitters. *Ann. Rev. Pharmacol. Toxicol.* **21,** 165–204.

Wood J. H., Hare T. A., Galeser B. S., Ballenger J. C., and Post R. M. (1979) Low cerebrospinal fluid γ-aminobutyric acid content in seizure patients. *Neurology* **29,** 1203–1208.

Wurtman, R. J., Hefti, F., and Melamed E. (1981) Precursor control of neurotransmitter synthesis. *Pharmacol. Rev.* **32,** 315–335.

Yamamoto S., Kiyama S., Watanabe Y., and Makita M. (1982) Practical gas-liquid chromatographic method for the determination of amino acids in human serum. *J. Chromatogr.* **233,** 39–50.

Chapter 3

Gas Chromatography–Mass Fragmentography of Amino Acids

PAUL L. WOOD AND D. L. CHENEY

1. Introduction

The coupling of a gas chromatograph (GC) to a mass spectrometer (MS) results in the marriage of two powerful technologies. The tremendous separative power of the GC is coupled to the *direct* identification and quantitation capabilities of the MS. The MS produces ions by electron bombardment of eluted compounds and the subsequent decomposition of these ions is monitored. The observed fragmentation patterns or "mass spectra" have characteristic ions that can be selectively monitored to obtain enhanced sensitivity for quantitative studies. This selected ion monitoring (SIM) is most accurately termed "gas chromatography–mass fragmentography (GC–MF)."

In the neurochemical analyses of amino acids, GC–MF is presently the most versatile methodology for the simultaneous separation, identification, and quantitation of amino acids. The following characteristics are inherent:

1.1. Sensitivity

Femtomole sensitivity is routine and is dependent upon the derivatization procedure and GC–MF mode of analysis.

1.2. Specificity

In GC–MF, a structurally specific ion with a defined GC retention time, is monitored, making this the most specific physiochemical methodology for amino acid analysis. In addition, several characteristic ions can be monitored and their ratios for a biological sample compared with those of the corresponding standards, if the identity of a biological peak is in question. No such control is possible with other methodologies.

1.3. Accuracy

To control for transfer losses, for variation in extraction, and for inequalities in derivative formation, it is now standard practice to add internal standards, labeled with a stable isotope, to the biological sample being analyzed. Such internal standards are chemically identical to the endogenous biological substance except that their molecular weight is greater than the biological sample, thereby avoiding interferences by natural isotopic content (Gaffney et al., 1971). Structural analogs also have been advocated as internal standards, but the assay variability is much greater, thereby reducing the accuracy of such measurements.

1.4. Speed

Sample preparation for most GC–MF analyses of amino acids is very simple and rapid. One hundred samples easily can be prepared daily. Furthermore, GC-run times for amino acid analyses are very rapid; for example, analysis of glutamate, gamma-aminobutyric acid (GABA), and aspartate requires less than three minutes. Clearly, GC-MS systems have a high capacity for amino acid measurements.

1.5. Ease of Column Maintenance

In contrast to the expensive and short-lived high pressure liquid chromatography (HPLC) columns used for amino acid analyses, GC columns are extremely easy to maintain and very inexpensive to repack.

1.6. Utility for Measurements of Amino Acid Dynamics

As with other neurotransmitter systems, simple measurements of steady-state levels of amino acids are of limited value. A much more valuable measure is the rate of utilization of the amino aid of interest. With GC–MF, it is possible to label an amino acid pool with stable isotopes and subsequently monitor, with one GC injection, the endogenous amino acid level, the amount of stable isotope incorporated into the amino acid pool, and a stable isotope internal standard to correct for sample recovery.

2. Purification Procedures

Purification of amino acids is required under circumstances in which maximum sensitivity is required or when interfering compounds are present. The total amino acid fraction may be isolated by ion-exchange chromatography with Dowex 50W-X8 (Moroni et al., 1980; Moroni et al., 1981). In other derivatization procedures, asparagine is hydrolyzed to aspartate and, similiarly, glutamine to glutamate. In these cases, selective isolation of glutamate and aspartate with Dowex AG 1-X8 (Corradetti et al., 1983) is preferred.

3. Derivatization

There are a large number of derivatization procedures for amino acids; however, three derivatives are predominantly utilized in biological studies (Table 1, Fig. 1).

TABLE 1
Useful EI Fragments for SIM of Amino Acid Derivatives

Derivatives	Amino acid	Mol wt	m/z
TAB	Gly	227	126,228
	Asp	341	240,342
	Glu	355	198,207
	GABA	255	232,204
PAB	Glu	405	304,248
	GABA	305	232,204
PAM	Asp	307	248,206
	Glu	321	262,202
	GABA	263	232,176
PAH	Gly	371	176,224
	Asp	579	216,384
	Glu	593	230,398
	GABA	399	176,232
TMS	Gly	291	174,276
	Asp	349	232,218
	Glu	363	246,348
	GABA	319	174,304

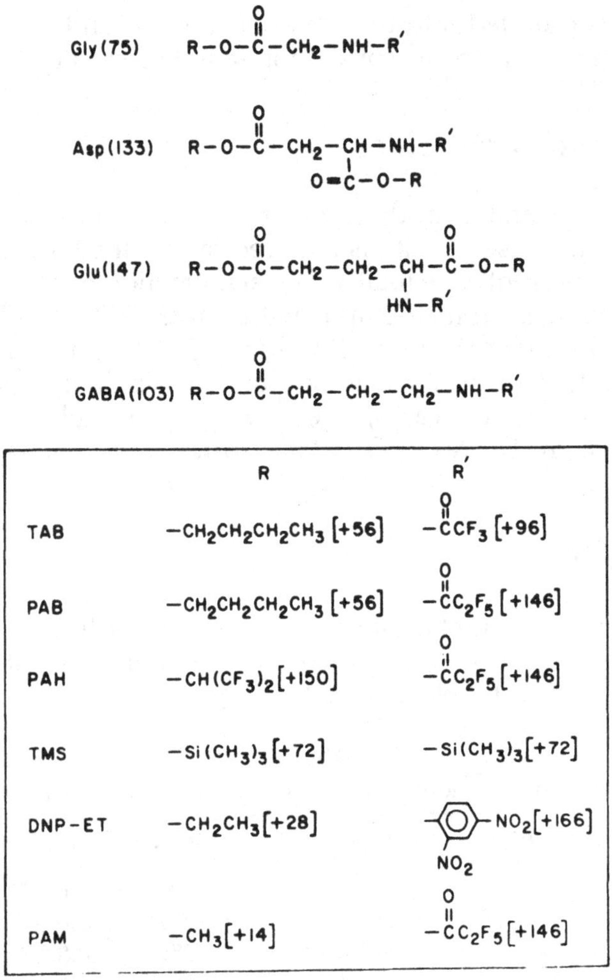

Gly (75) $R-O-\overset{\overset{O}{\|}}{C}-CH_2-NH-R'$

Asp (133) $R-O-\overset{\overset{O}{\|}}{C}-CH_2-\underset{\underset{O=C-O-R}{|}}{CH}-NH-R'$

Glu (147) $R-O-\overset{\overset{O}{\|}}{C}-CH_2-CH_2-\underset{\underset{HN-R'}{|}}{CH}-\overset{\overset{O}{\|}}{C}-O-R$

GABA (103) $R-O-\overset{\overset{O}{\|}}{C}-CH_2-CH_2-CH_2-NH-R'$

	R	R'
TAB	$-CH_2CH_2CH_2CH_3$ [+56]	$-\overset{\overset{O}{\|}}{C}CF_3$ [+96]
PAB	$-CH_2CH_2CH_2CH_3$ [+56]	$-\overset{\overset{O}{\|}}{C}C_2F_5$ [+146]
PAH	$-CH(CF_3)_2$ [+150]	$-\overset{\overset{O}{\|}}{C}C_2F_5$ [+146]
TMS	$-Si(CH_3)_3$ [+72]	$-Si(CH_3)_3$ [+72]
DNP-ET	$-CH_2CH_3$ [+28]	$-\langle O \rangle-NO_2$ [+166] NO$_2$
PAM	$-CH_3$ [+14]	$-\overset{\overset{O}{\|}}{C}C_2F_5$ [+146]

Fig. 1. Derivatives for GC–MF analysis of glycine, aspartate, glutamate, and GABA. To obtain the molecular weight of the derivative, add to the molecular weight of the amino acid the molecular weight of the derivative fragments that are in brackets. For example, PAH-Glu [147 + 2(150) + 146 = 593]. Note that for TMS–GABA both hydrogens of the amino group are replaced.

3.1. PFPA–HFIP (PAH)

In this procedure, the amino groups are acylated with pentafluoropropionic anhydride (PFPA) and the carboxylic groups esterified with hexafluroisopropanol (HFIP) in a one-step procedure (Bertilsson and Costa, 1976). Typically, 100 μL of PFPA and 50 μL of HFIP are added to a freeze-dried or nitrogen-dried tissue extract and the mixture heated at 70°C for 1 h. The samples are cooled, dried under nitrogen to remove the excess reagents that are volatile, and then dissolved in ethyl acetate for injection into the GC–MS. This derivative has been used extensively for measurements of central nervous system (CNS) glutamate, glutamine, and GABA (Moroni et al., 1980; Wolfensberger et al., 1982 Dessort, 1982), as well as glycine (Lapin and Karobath, 1980).

Since this reaction goes essentially to completion, it is useful for the analysis of CNS tissue extracts and has more than adequate sensitivity for analysis of tissue micropunches (1–2 mg tissue). For biological fluids and perfusates, some sample cleanup is generally required before using this reaction. The most convenient cleanup procedure appears to be the use of Dowex 50W-X8 resin. The amino acids are retained during washing procedures and subsequently eluted with 3N NH₄OH (Moroni et al., 1980, 1981, 1983). The eluate can be dried and then reacted with PFPA and HFIP. For perfusates from release experiments, sample cleanup may not be an absolute requirement as salts do not appear to inhibit the acylation (Murayama et al., 1981). The use of glass beads (80–120 mesh) to catalyze the reactions with amino acids from perfusates has also been described (Wolfensberger et al., 1979, 1981, 1982).

These derivatives have been excellent for analyses of amino acids from neuronal tissues (Fig. 2). However, to analyze the trace levels of amino acids present in either cerebrospinal fluid or perfusates from release studies, more stable derivatives are preferred.

3.2. N-Acyl Alcohol Esters

The derivatization of carboxyl groups with HFIP has limitations because the derivatives are unstable in dilute solutions and are prone to extensive fragmentation even in chemical ionization (CI) mode. Therefore, for analyses of trace amounts of amino acids, particularly in biological fluids or perfusates, the alcohol esters are preferred. In most cases, these derivatives, even in dilute solutions, are stable for 1–2 wk at room temperature.

The esters are prepared by an initial acid-catalyzed esterification in an anhydrous alcohol. The most useful alcohols are *n*-butanol and methanol. In the case of 3N HCl–*n*-butanol, the acidified reagent can be obtained commercially from Regis Chemicals, IL. The most popular methylating reagents are 10% boron trichloride in methanol that can be obtained from Sigma Chemicals, MO; 20% HCl in methanol (dropwise addition of 1 mL of acetyl chloride to 4 mL of methanol; Corradetti et al., 1983); and thionyl chloride in methanol (dropwise addition of 1 mL of thionyl

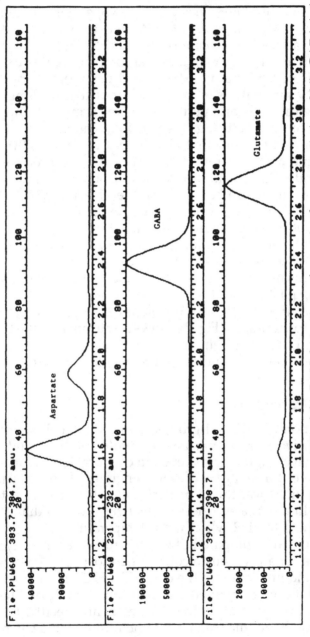

Fig. 2. GC–MS scan (EI) of the PAH derivatives of rat striatal aspartate (m/z = 384), GABA (m/z = 232), and glutamate (m/z = 398). Tissue was extracted with 1N HCl, freeze-dried, and reacted with PFPA/HFIP prior to analysis. Column was 3% SP 2250 (80/100; 6 ft); Temperature 1 (T_1) = 140; followed by a 30°C/min gradient.

chloride to 9 mL of methanol; Faull et al., 1978). The esterification with methanol can be performed at 85°C for 15 min or overnight at 40°C whereas esterification with *n*-butanol is best driven at 120°C for 30 min to obtain good derivatization of glutamate and asparate (MacKenzie and Tenaschuk, 1979a,b).

The subsequent acylation step is generally performed with trifluoroacetic anhydride (TFAA), PFPA, or heptafluorobutyric anhydride (HFBA), all of which yield excellent derivatives for GC–MS. The most popular derivatives include: (a) *TAB* (TFA/*n*-BuOH): Summons et al., 1974; Schulman and Abramson, 1975; Petty et al., 1976; Kingston and Duffield, 1978; Finlayson et al., 1980; (b) *PAB* (PFPA/*n*-BuOH): Sjoquist, 1979; Lapidot and Nissim, 1980; (c) *TAM* (TFA/MeOH): Zaylak et al., 1977; Coutts et al., 1979; (d) *PAM* (PFPA/MeOH): Faull et al., 1978; Ferkany et al., 1978; Corradetti et al, 1983; (e) *HAM* (HFBA/MeOH): Faull and Barchas, 1983; and (f) *HAB* (HFPB/*n*-BuOH): Lapidot and Nissim, 1980; Bengtsson et al., 1981.

Of these derivatization procedures, the one that we have found the most sensitive for dilute samples is the PAM procedure. However, as with all the procedures utilizing alcohol esters, this one has a drawback. During the esterification reaction, glutamine is hydrolyzed to glutamate and asparagine to aspartate. Therefore, these dicarboxylic acids must be isolated from the carboxamides prior to esterification. Ion exchange on Dowex AG1-X8 is extremely useful for this procedure (Corradetti et al., 1983):

    a. Resin preparation: Dowex AG1-X8 (200–400 mesh, chloride form) is washed six times in distilled water, with the light debris being decanted with each wash. The resin is next washed twice in 2*N* HCL and another three times in distilled water. The resin is stored under water.

    b. Column preparation: A 5.75-in. Pasteur pipet is plugged with silanized glass wool and a 1 in. resin bed applied. This resin is washed with 10 mL of water.

    c. The sample and internal standard are dissolved in borate buffer (pH 8.9) and loaded onto the column.

    d. The column is washed with 10 mL of water (wash contains GABA and glutamine).

    e. The column is next washed with 1 mL of 0.02*N* HCL.

    f. The dicarboxylic amino acids (glutamate and aspartate) are eluted with 0.75 mL of 1*N* HCl.

g. The sample is dried in a 1 mL reactivial (Pierce Chemicals) using a Savant Vac Torr Evaporator.

h. One hundred μL of 10% boron trichloride/methanol (Sigma) is added to the dry sample and heated at 40°C overnight.

i. The sample is next dried under N_2.

j. Fifty μL of ethyl acetate and 100 μL of PFPA are added to the *dry* samples prior to heating at 70°C for 60–90 min.

k. The samples are cooled and dried under N_2.

l. Ethyl acetate (20 μL is added and an aliquot injected into the GC–MS (Figure 3):

   (i) Column: 6% QF-1 on Suplecoport (100/120) Temperature program: 180°C for 1 min, increasing at 20°C/min to 210°C Helium = 25 cc/min

   (ii) [Figure 3 near here]

Masses for EI and CI SIM

	GABA (d_6)	ASP (d_3)	Glu (d_5)
m/z(EI)	232,204 (238)	248,216 (215)	262,230 (267)
m/z (NCI)	243 (249)	287 (290)	301 (306)
RT (min)	1.45	1.55	2.4

3.3. Trimethylsilyl (TMS) Derivatives

The analysis of a large number of neurotransmitters and their metabolites as TMS derivatives has been excellently reviewed by Abramson and coworkers (1974). The electron impact (EI) spectra of all the TMS-amino acids have also been reviewed (Iwase et al., 1979). In neurochemical studies, a number of workers have used the TMS derivatization procedure to measure tissue GABA (Cattabeni et al., 1976, 1977; Holdiness et al., 1981; Haseyawa et al., 1981). TMS-glutamate has been found to be extremely valuable for GC–MS analysis in the CI mode with ammonia as the reagent gas (Dessort et al., 1982). An example of the use of TMS derivatives to monitor glutamate and aspartate release in brain dialysis experiments is presented in Fig. 4.

4. Assay Parameters

4.1. Precision and Accuracy

Internal standards labeled with a stable isotope offer tremendous analytical advantages because they optimize the precision and ac-

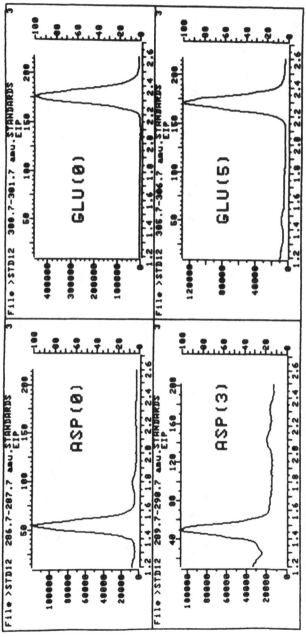

Fig. 3. GC–MF analysis (NCI, methane reagent gas) of the PAM derivatives of glutamate and aspartate; conditions as in 12a (see text).

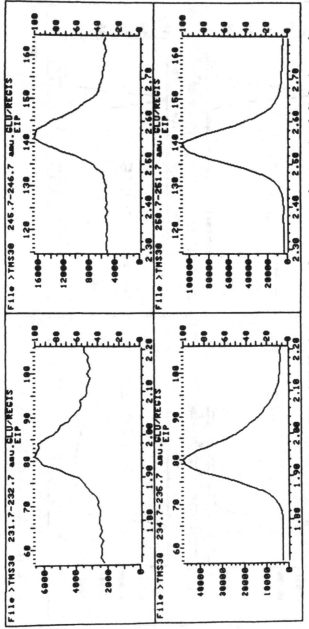

Fig. 4. GC–MF analysis (EI) of TMS glutamate and aspartate from striatal dialysis perfusates. Perfusates were vacuum dried and reacted with Sylon BFT (Supelco)/pyridine (1:1) at 60°C for 3 h. Column was 3% SP 2250 (6 ft); $T_1 = 180°C$.

curacy of SIM analyses. They correct for variations in (1) extractions, (2) chemical reactions, (3) transfer steps, (4) evaporative losses during drying steps, (5) oxidative losses, (6) "on column" losses, and (7) instrument or column drift. In cases where interfering peaks are present, these internal standards also serve as internal markers since they have identical retention times with those of the endogenous amino acids.

The amount of internal standard to add to a biological sample has been a matter of some debate. Many workers have added a large excess in the hopes that this would act as a carrier. However, isotopic impurities can in such cases cause considerable "crosstalk" into the SIM channel of the biological sample. A more ideal approach is to add the identical amount of internal standard as is present in the biological sample under normal conditions. With this procedure, decreases and increases in the endogenous amino acid level can be accurately monitored. An additional feature of this approach is that a full standard curve need not always be constructed since with deuterated internal standards a wide dynamic range in such curves is observed. Indeed, mole ratios (endogenous/internal standard) are linear up to 100 and down to 0.1 (Colby and McCamen, 1979).

4.2. Limit of Detection

Internal standards are effective controls for assay precision and accuracy, but many more variables determine the limit of detection for a given assay. To obtain the best signal-to-noise ratio, the key areas of concern include:

4.2.1. Sample Preparation

Sample cleanup procedures can reduce biological background in cases of trace analyses. However, cleanup procedures utilizing impure ion exchange resins and impure organic solvents may actually add contamination to the analytical sample.

4.2.2. GC Column

The choice of a GC column is dependent upon minimizing column bleed; separation of the amino acids of interest from interfering peaks and background; rapid analysis time; and minimal sample adsorption. The columns utilized at the present time are presented in Table 2.

4.2.3. SIM Scanning Parameters

During a SIM run, the proper selection of the electron multiplier gain determines to a great extent the ultimate sensitivity for the analysis. The dwell time for each ion is also of importance. Ten to

TABLE 2
GC Columns[a] Used for GC–MF Analyses of Amino Acids

Derivatives	Column phase	Advantage/ disadvantage	Reference
TAB	Tabsorb (0.25% EGA)	Excellent separations	Pettey et al., 1976 Kingston and Duffield, 1978
PAB	1% OV-17	Fast run times	Sjoquist, 1979 Lapidot and Nissim, 1980
PAM	3% SE 54	Good separations	Faull et al., 1978
	3% SP2250	Low background	Corradetti et al., 1983
	6% QF-1	Low background	Wood (unpublished)
PAH	3% OV-17	Fast run time Gln lost on column	Bertilsson and Costa, 1976
	6% OV-17	Excellent separations	Huizinga et al., 1978
	2% QF-1	Low background	Saunder et al., 1981 Dessort et al., 1982
	3% QF-1	Low background Separation of Glu and Gln	Wolfensberger et al., 1979
	6% QF-1	Separation of Glu and Gln	Wolfensberger et al., 1981, 1982
	3% OV-1	Separation of Glu and Gln	Moroni et al., 1980
TMS	3% OV-17	Good separations	Cattabeni et al., 1976 Holdiness et al., 1981
	3% OV-11	Good separations	Haymond et al., 1980
	1.5% OV-101	Fast run time	Iwase et al., 1979
	2% QF-1	Low background	Dessort et al., 1982
	3% Dexsil 300	Excellent separations	Abramson et al., 1974

[a]Supports include Chromosorb WHP, Gas Chrom Q, and Supelcoport.

20 scans of each ion for a given GC peak are required (Matthews and Hayes, 1976; Falkner, 1981). This generally means a dwell time of 50–250 ms with undirectional scan cycles (mass 1, mass 2, mass 3; mass 1, mass 2, mass 3; . . .) being preferred (Matthews and Hayes, 1976). Wherever possible, the scanning of high mass (m/z) ions is also better since it reduces interferences from background contamination.

4.2.4. MS Mode

With computer tuning of mass spectrometers, the interchangeable use of EI, positive chemical ionization (PCI) and negative chemical ionization (NCI) mass fragmentography can be greatly facilitated. In our laboratory, using a Hewlett Packard 5987 GC–MS, it is possible to establish parameters for all three modes and save them on a disc file within 15–30 min. Subsequently, any of the modes can be utilized by calling up the tuning file during the day. This approach is extremely valuable when new assays are being developed and all three modes are under evaluation for best sensitivity.

The most common mode for analysis of amino acids has been EI. However, with all the described derivatives, considerable fragmentation occurs in this mode. Similiarly, PCI with methane leads to extensive fragmentation of TAB derivatives (Petty et al., 1976), with no enhancement in sensitivity (Holdiness et al., 1981); however, with isobutane, MH^+ ions predominate (Petty et al., 1976; Finlayson et al., 1980). In the case of ammonia PCI, an excellent molecular ion of PAH-GABA ($MH^+ + NH_3 = 417$) is evident. This is clearly the most useful method for measuring CNS GABA levels; however, other PAH amino acids do not form stable ions under these conditions. Dessort et al. (1982), have also observed this problem with the PAH derivative of glutamate; however, in this case TMS-glutamate forms an excellent ion under ammonia PCI ($MH^+ = 364$).

With the PAM derivatives of amino acids, EI has been used to monitor the release of glutamate, aspartate, and GABA (Corradetti et al., 1983). Methane CI also has been used to monitor higher molecular weight ions ($m/z = 264$) in measurement of blood GABA (Ferkany et al., 1978). Further improvements in the assay of these amino acids occur when they are measured in NCI (Fig. 5). In this case GABA, glutamate, and aspartate all form key M-HF anions when derivatized with PFPA (see Wood, 1982). The advantages of this assay are the low background and the great sensitivity. In addition, since no fragmentation of the parent amino acids takes place, this assay is excellent for monitoring the incorporation of ^{13}C-glucose into amino acid pools. Similar increases in sensitivity with the NCI analysis of the HAM derivative of GABA have been reported (Faull and Barchas, 1983).

Another important example of the importance of the GC–MS mode in analytical studies is with quinolinic acid. The methylated derivative of this tryptophan metabolite undergoes extensive fragmentation in EI and is, therefore, not useful for the assay of

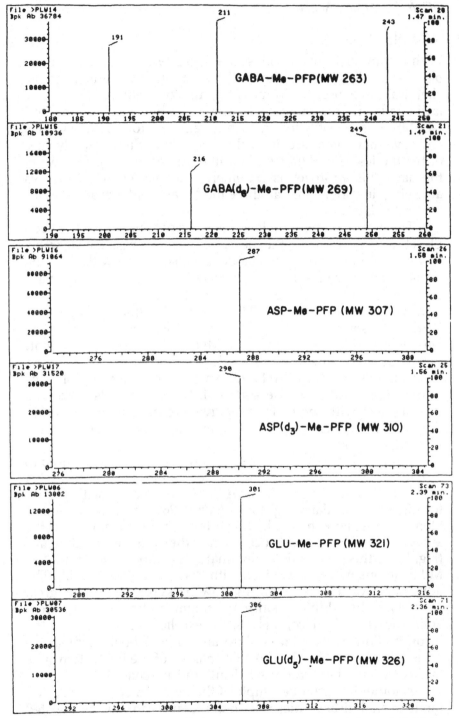

Fig. 5. Mass spectra (NCI) of the PAM derivatives of GABA, glutamate, and aspartate. See Corradetti et al. (1983) for the comparable EI spectra.

biological samples. However, in PCI the $(M + H)^+$ cation predominates and is excellent for SIM studies (Fig. 6).

5. Enzyme Assays

The use of GC–MF to measure the activity of glutamic acid decarboxylase (GAD; EC 4.1.1.15) has been reported, with the conversion of glutamate-d_5 (Cattabeni et al., 1977) or glutamate-d_0 (Holdiness et al., 1981) to GABA being monitored.

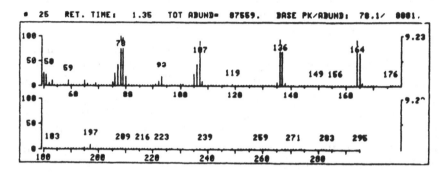

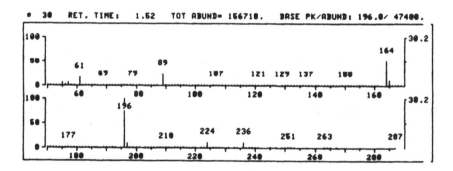

Fig. 6. EI (upper) and PCI (lower; methane reagent gas) of the methylated derivative of quinolinic acid. The degree of fragmentation in PCI mode is much reduced, with the molecular ion predominating.

This approach offers advantages over measurements of CO_2 evolution that can also be generated by other mechanisms.

The activity of phenylalanine-4-monooxygenase (EC.1.14. 16.1) has been monitored in vivo in man by monitoring the conversion of phenylalanine-d_5 to tyrosine-d_4 (Zaylak et al., 1977).

6. Biological Fluids

The endogenous levels of amino acids in blood and urine have been measured using GC–MF (Schulman et al., 1975; Petty et al., 1976; Mee et al., 1977; Kingston et al., 1978; Irving et al., 1978; Finlayson, 1980).

As a result of the ethical problems associated with the clinical use of radioisotopes, the use of amino acids labeled with stable isotopes also has made great inroads in clinical studies (Halliday and Rennie, 1982). Such studies have been able to define the level of amino acid transamination and oral bioavailability in man (Irving et al., 1978; Haymond et al., 1980; Matthews et al., 1981a). In addition, the use of dual labels (e.g., [^{15}N, ^{13}C-leucine) also has enabled the simultaneous evaluation of amino acid transamination and decarboxylation (Matthews et al., 1981b).

7. CNS Levels

Amino acids probably act as neurotransmitters at a majority of all synapses in the CNS (Bloom and Iversen, 1971; Krnjevic, 1974). In addition to their role as putative neurotransmitters, these compounds occupy central positions in the general metabolism (Johnson, 1972). It is difficult, if not impossible, to differentiate the amino acid pools related to the metabolic function from the amino acid pools related to the neurotransmitter function. Because of their multiplicity of roles, it is not surprising that the amino acids are widely distributed throughout the CNS in fairly high concentrations (Table 3).

7.1. Microwave Fixation and Brain Amino Acid Levels

It has been known for many years that GABA levels in the brain rise rapidly after death (Baxter, 1972; Lovell et al., 1963; Shaw and Heine, 1965; Minard and Mushahwar, 1966; Shank and Aprison 1971). Levels of other amino acids are modified postmortem, but not to the same extent as GABA (Perry et al., 1981). Decapitation into liquid nitrogen prevents the postmortem rise in GABA; how-

TABLE 3
Concentrations of Amino Acids in Various Rat Brain Areas

Brain area	GABA, nmol/mg protein	Glu, nmol/mg protein	Asp, nmol/mg protein	Gly, nmol/mg protein	Gln, nmol/mg protein
Mouse					
Whole brain	14.3 ± 0.5[a]	—	—	—	—
	18 ± 1.0[f]				
Rat					
Septum	52 ± 1.4[b]	106 ± 2.0[b]	43 ± 7.0[b]	14 ± 2.0[b]	—
	87 ± 8.7[d]	160 ± 3.1[d]			
	35 ± 1.4[g]	160 ± 8.0[h]			
	60 ± 2.0[h]				
N. accumbens	100 ± 4.8[d]	150 ± 10[d]	28 ± 6.0	10 ± 2.0[h]	—
	47 ± 2.2[c]	125 ± 3.4[c]			
	46 ± 2.1[g]	125 ± 10[h]			
	28 ± 3.0[h]				
Cortex					
frontal	40 ± 2.1[d]	132 ± 3.4[d]	26 ± 0.6[c]	12 ± 0.7[c]	41.0 ± 2.1[c]
	51 ± 1.9[c]	70 ± 2.5[c]	42 ± 8.0[h]	11 ± 1.0[h]	
	17 ± 1.5[h]	130 ± 7.0[h]			
Parietal	37 ± 1.8[d]	120 ± 9.9[d]	46 ± 10[h]	14 ± 3.0[h]	—
	48 ± 10[h]	118 ± 14[h]			
Occipital	32 ± 2.2[h]	121 ± 7.0[d]	52 ± 10[h]	15 ± 2.0[h]	—
	19 ± 3.0[h]	108 ± 10[h]			
Temporal	26 ± 1.3[d]	110 ± 5.2[d]	33 ± 5.0[h]	15 ± 3.0[h]	—
	38 ± 50[h]	120 ± 15[h]			
Cerebellum	15 ± 0.5[c]	110 ± 3.0[c]	—	—	49 ± 4.9[c]
	13 ± 1.9[g]				
Globus pallidus	62 ± 2.6[d]	80 ± 4.7[d]	35 ± 5.0[h]	13 ± 3.0[h]	58 ± 6.0[c]
	89 ± 5.5[c]	95 ± 5.0[c]			
	80 ± 3.0[h]	87 ± 15[h]			
Hippocampus	21 ± 2.0[c]	93 ± 4.0[c]	—	—	51 ± 7.2[c]
	60 ± 3.0[d]	200 ± 8.8[d]	25 ± 0.7[c]	9.7 ± 0.3[c]	51 ± 1.5[c]
	27 ± 1.5[c]	104 ± 2.8[c]	33 ± 8.0[h]	8.0 ± 1.0[h]	
	20 ± 0.5[g]	86 ± 7.0[h]			
	21 ± 2.0[h]				
N. caudatus	46 ± 0.3[d]	130 ± 2.9[d]	31 ± 2.0[h]	5.0 ± 1.0[h]	66 ± 2.7[c]
	20 ± 3.0[c]	81 ± 4.0[c]			
	22 ± 0.8[g]	102 ± 12[h]			
	30 ± 5.0[h]				
Substantia nigra	95 ± 4.4[c]	79 ± 2.8[c]	27 ± 4.0[h]	12 ± 3.0[h]	72 ± 7.0[c]
	70 ± 2.1[d]	88 ± 2.2[d]			
	42 ± 5.2[g]	60 ± 10[h]			
	62 ± 6.0[h]				
Hypothalamus	38 ± 1.5[g]	63 ± 10[h]	23 ± 3.0[h]	11 ± 1.5	—
	33 ± 2.0[h]				

[a]Richardson and Scudder (1976), enzymatic method, whole body microwave.
[b]Cheney (1984), GC–MF, focused microwave.
[c]Costa et al. (1979), GC–MF, focused microwave.
[d]Revuelta et al. (1981), GC–MF, focused microwave.
[e]Hassler et al (1982), amino acid analyzer, decapitation.
[f]Haseyawa et al. (1981), GC–MF, focused microwave.
[g]Balcom et al. (1975), enzymatic method, focused microwave.
[h]Lane et al. (1982), chromatography of ^{3}H-dinitroflurobenzene derivatives, quick freezing.

ever, detailed dissection of the frozen brain is extremely difficult. Microwave irradiation has become the preferred method of sacrifice since the rapid, uniform heating of the tissue denatures the enzyme systems responsible for postmortem changes (Guidotti et al., 1974; Balcom et al., 1975; Richardson and Scudder, 1976). Rapid elevation of the brain core temperature to a level sufficient to eliminate enzyme activity only can be achieved if the microwave beam is of sufficient intensity and is properly focused (Guidotti et al., 1974).

7.2. Extraction of Amino Acids

Various techniques have been used to extract amino acids from biological tissues prior to analysis by GC–MF. Unfortunately, there have been no systematic studies comparing the various procedures. The methods currently being used include extraction by trichloroacetic acid, perchloric acid, sodium acetate, zinc sulfate, ethyl alcohol, and hydrochloric acid.

7.2.1. TCA Extraction

Amino acids can be extracted from biological tissue or tissue powder with 5–7% trichloroacetic acid (Freeman et al., 1980; Berl and Frigyesi, 1969). The trichloroacetic acid has to be removed prior to preparation for GC–MF analysis by successive washes with ether and the washed extract reduced to dryness.

7.2.2. Perchloric Acid Extraction

Biological tissues can be homogenized in 0.4–$0.5N$ perchloric acid and the supernatants neutralized with $KHCO_3$ or KOH to precipitate the perchlorate (Okada et al., 1971; Balcom et al., 1975). Prior to GC–MF, the supernatants must be freeze dried or evaporated under nitrogen.

7.2.3. Sodium Acetate Extraction

Biological tissues can be homogenized in $0.05M$ sodium acetate (pH 6.1) (Balcom et al., 1975) and the supernatants reduced to dryness.

7.2.4. Zinc Sulfate Extraction

Biological tissue can be homogenized in $0.3N$ $ZnSO_4$ in the proportions of 0.6 g of tissue to 1 mL of reagent. Barium hydroxide (1 mL of $0.3N$) is then added and the mixture is again homogenized and centrifuged. The use of barium instead of sodium hydroxide to neutralize the zinc sulfate results in an extract free of sulfate ions, but may be attended by additional loss of some constituents by absorption on the barium sulfate precipitate (Blass, 1960). The supernatant must be reduced to dryness prior to reaction and analysis by GC–MF.

7.2.5. Ethyl Alcohol Extraction

Biological tissue can be homogenized in 80% aqueous ethanol. The supernatant may (Haseyawa et al., 1981) or may not (Bertilsson et al., 1977; Balcom et al., 1975) be extracted with hexane for further purification. Prior to analysis by GC–MF the aqueous phase must be reduced to dryness.

7.2.6. Hydrochloric Acid Extraction

Biological tissue can be homogenized in $1.0N$ HCl and the supernatant reduced to dryness by freeze-drying (Robitalle et al., 1982).

In each of the extraction procedures, it is necessary to reduce the samples to dryness either under nitrogen or by vacuum centrifugation without concentrating the acid or without increasing the salt concentration. Balcom et al., (1975) has compared the levels of GABA in hemibrains of microwaved rats using the ethanol extraction (Baxter and Robert, 1959), the perchloric acid extraction (Okada et al., 1971) and the sodium acetate extraction procedure described above. Ethanol (80%) appears to be more effective than $0.05M$ sodium acetate which, in turn, is more effective than $0.5N$ perchloric acid in extracting GABA. In ethanol extraction procedures (Balcom et. al, 1975) sonication and homogenization are equally effective as techniques for tissue disruption.

8. Amino Acid Dynamics

During the last few years, the focus of amino acid research has shifted from measurements of steady-state concentrations to the investigation of the dynamic state of the neuronal and nonneuronal stores. In the past, investigators have asked how environment, neuronal activity, or drugs increase or decrease the concentration of amino acids in brain structures. Now it is becoming more commonplace for investigators to ask how drugs and environment affect the neuronal function without changing the steady state of the amino acids. It is presently believed that changes in neuronal activity are paralleled by a modification of the dynamic state of the individual amino acids that may, or may not, be accompanied by changes in the amino acid concentration. Biochemical methods developed for the simultaneous quantitation by GC–MF of various amino acids have made it possible to begin to answer some of the questions about the dynamic relationship between the amino acids and their putative precursors.

Recently, Lane et al. (1982) have evaluated the role of amino acid neurotransmitters in an animal model of anxiety and have been able to demonstrate very few changes in the content of amino acids in twenty rat brain areas. Such results suggest several options: (1) amino acids may be unimportant in animal models of anxiety; (2) behavioral manipulations may affect small physiologically functional pools of amino acid neurotransmitters whose fluctuations are not readily detectable by gross measurement of amino acid pools; or (3) physiologically relevant modifications in behavior may not be expected to modify steady-state kinetics, but may affect the rate of utilization of the amino acids. In further evaluations of amino acid dynamics, Lane et al. (1982) have been able to demonstrate numerous changes in the utilization of various amino acid neurotransmitters, emphasizing the utility of dynamic measurements over tissue content alone.

8.1. GABA Dynamics

GC–MF quantitation of the levels of GABA and its precursor glutamate in brain tissue (Bertilsson and Costa, 1976) has made it possible to evaluate the dynamic relationship between glutamate and GABA in small brain structures (Bertilsson et al., 1977; Mao et al., 1978). Only isotopic methodology allows the study of biochemical events in the intact living organism without disturbing steady-state processes. This has been done by monitoring the changes with time in the enrichment of ^{13}C in glutamate and GABA in various brain nuclei following constant rate infusions of ^{13}C-glucose (Fig. 7).

Although glucose is separated metabolically from glutamate and GABA by numerous steps, it readily crosses the blood-brain barrier and rapidly labels both the precursor glutamate and the product GABA. From the relationship between the precursor and product curves, GABA utilization has been calculated by assuming that immediately after infusion of ^{13}C-glucose the neuronal GABA that is selectively labeled is stored in a single open metabolic compartment. Thus, the equations derived by Racagni et al. (1974) have been used to calculate the fractional rate constant for GABA efflux (k_{GABA}). This value multiplied by the steady-state concentration of GABA gives an estimate of the rate of utilization of GABA. This measurement of GABA utilization is useful for comparative studies of drug or transmitter interactions, but fails to yield absolute values of GABA synthesis rates (Cheney, 1984).

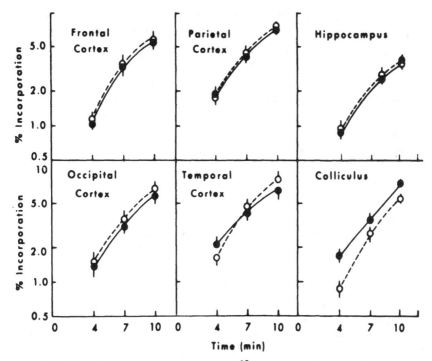

Fig. 7.   Percent incorporation of ^{13}C-glucose (75 μmol/kg/min; 0.1 mL/min) into glutamate (closed circles) and GABA (open circles) of various brain nuclei as a function of time (Revuelta et al., 1981; reproduced with permission).

This calculation yields reliable information about neurotransmitter dynamics in nuclei that contain terminals of long axon GABA neurons or in nuclei that contain a population of GABAergic interneurons that do not synapse with each other. However, in brain nuclei containing GABA interneurons that are interconnected, the utilization rate of GABA gives average or composite data about the rate of activity. The mathematical model described by Racagni et al. (1974) implies that the glutamate pool (labeled from glucose at a fast rate) serves as a precursor of GABA. This assumption is never completely true, since glutamate functions not only as a GABA precursor, but also as an intermediate metabolite and a primary transmitter. Because of the difficulty of defining the precursor pool of glutamate, the calculated rates of utilization are not absolute values but relative values. If the pool of glutamate serving as a precursor of transmitter GABA is labeled at a fast rate, then the method would tend to overestimate the actual rate of utilization. By collecting the samples very early during the infusion of ^{13}C-glucose and by selecting those tissues in which the pool of glutamate that functions predominately as a GABA precursor prevails over that of the other two glutamate pools, the errors derived from such a generalization can be minimized. Thus, the rate of GABA utilization has been estimated in nucleus accumbens, nucleus caudatus, globus pallidus, substantia nigra, tuberculum olfactorium, septum and colliculus (Table 4) when the rate of change of ^{13}C incorporation into glutamate and GABA fulfills some minimal criteria for expression of a precursor–product relationship (Fig. 7; Revuelta et al., 1981). Since these criteria are not fulfilled in cerebellar cortex, deep cerebellar nuclei, hippocampus, or in areas of the cortex, dynamic measurements by GC–MF have not been determined in these areas (Fig. 8; Revuelta et al., 1981). As shown in Fig. 8, the curves for the incorporation of ^{13}C are superimposed on the curves for the incorporation of label into glutamate, making it impossible to determine the rates of utilization in these areas.

A question that must be answered when conducting isotope studies is whether or not the utilization rate calculated following constant infusion of a stable isotopic precursor can be used as an index of neuronal activity. In the case of the striatonigral GABAergic pathway, electrical stimulation and hemitranssection experiments have demonstrated that the utilization rate of GABA is proportional to the activity of this pathway (Mao et al., 1978). However the tissue contents of GABA and glutamate do not correlate with the changes in neuronal activity. One week after hemitranssection, the contents of GABA and glutamate and the utilization rate for GABA are significantly reduced in the substan-

TABLE 4
GABA Turnover Rate in Various Regions of Rat
Brain[a]

Region	k_{GABA}, h^{-1}	GABA utilization, nmol/mg prot · h
N. accumbens	17 ± 2.5	1700 ± 240
N. caudatus	23 ± 2.0	1100 ± 82
G. pallidus	13 ± 1.4	790 ± 71
S. nigra	11 ± 2.1	1000 ± 58
T. olfactorium	25 ± 2.2	900 ± 48
Septum	16 ± 1.2	1500 ± 140
Colliculus	40 ± 3.2	1400 ± 85

[a]Data modified from Revuelta et al. (1981).

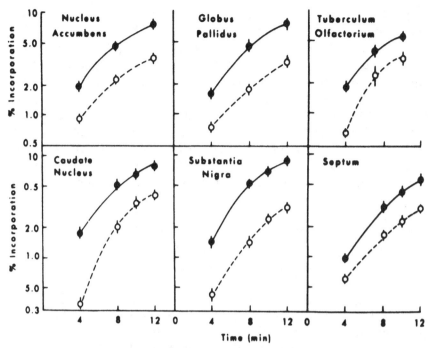

Fig. 8. Percent incorporation of ^{13}C-glucose into glutamate (closed circles) and GABA (open circles) of various brain regions as a function of time (Revuelta et al., 1981; reproduced with permission.)

tia nigra on the side of the lesion. Conversely, when the nucleus caudatus is stimulated electrically, there is a threefold increase in the utilization rate of GABA in substantia nigra without altering the concentration of GABA or glutamate in this nucleus (Mao et al., 1978). Thus the utilization rate of GABA in the nerve terminal region in the substantia nigra increases or decreases proportionally to the stimulation of the cell body region in the nucleus caudatus.

8.2. Glutamate Dynamics

Recently, Fonnum (1981) has stated, "There is a large body of literature concerning the synthesis and metabolism of glutamate . . . There are in the literature no studies pertaining to the turnover of the transmitter pool of glutamate only." Indeed, biochemical measurements directed to ascertain the dynamic state of the glutamate functioning as a transmitter must differentiate this pool from the other neuronal pool in which glutamate functions as a GABA precursor. Both pools must, in turn, be differentiated from the neuronal and glial cell pools when glutamate is a product of intermediary metabolism. If the precursors of glutamate were different in these various pools it would be possible to use rates of glutamate utilization to estimate participation of gluta-matergic neurons in brain function. Quastel (1974) has suggested that glutamine might function in replenishing the transmitter pool of glutamate. Glutamate is thought to function in replenishing the transmitter pool of GABA (Balazs et al., 1973). Several studies have described a large and a small compartment for glutamate (Balazs et al., 1973; Van den Berg et al., 1975; Fonnum, 1981). The precursors of glutamate appear to be different in these two compartments. Injection of labeled bicarbonate, acetate, propionate, butyrate, glutamate, aspartate, GABA, and leucine labels glutamine to a greater extent than glutamate. Injection of labeled glucose, pyruvate, lactate, glycerol, and hydroxybutyrate labels glutamate to a greater extent than glutamine (Berl and Clarke, 1960; Johnson, 1972). These results can be explained if the first group of precursors are preferentially metabolized in a small compartment of glutamate that is responsible for the glutamine synthesis. Fonnum (1981) agrees that glutamine is synthesized in the small compartment and converted to glutamate in the large compartment. With this precursor–product relationship between glutamine and glutamate, it may be possible to label the precursor glutamine preferentially and study the utilization rates of glutamate. However, a pulse-label injection of labeled acetate (Berl and Frigyesi, 1969) failed to produce a precursor–product relationship as defined by Zilversmit (1960). Moreover, attempts to obtain a precursor–product relationship between glutamine and glutamate by infusing deuterated glutamine into the ventricles have been disappointing (Costa et al., 1979).

The multiple pathways whereby the amino acid neuro-transmitters may be interconverted and the multiple compartments in which the amino acid neurotransmitters may be functional makes dynamic studies of glutamate, aspartate, and GABA a very difficult problem. So far, only GABA has been studied in detail.

9. Summary

GC–MF offers a number of sensitive and specific assays for neurotransmitter amino acids and also allows varied approaches to the measurement of the dynamics of these compounds. The safety of amino acids labeled with stable isotopes also makes such analyses extremely valuable for clinical studies.

References

Abramson F. P., McCamen M. W., and McCamen R. E. (1974) Femtomole levels of analysis of biogenic amines and amino acids using functional group mass spectrometry. *Anal. Biochem.* **51**, 482–499.

Balazs R., Patel A. J., and Richter D. (1973) Metabolic compartments in the brain; their properties and relation to morphological structures, in Metabolic Compartmentation in the Brain (Balazs R. and Cremer J. E., eds.), pp 167–185. Macmillan, London.

Balcom G. J., Lenox R. H., and Meyerhoff J. L. (1975) Regional γ-aminobutyric acid levels in rat brain determined after microwave fixation. *J. Neurochem.* **24**, 609–613.

Baxter C. F. (1972). Assay of γ-aminobutyric acid and enzymes involved in this metabolism, in Methods in Neurochemistry, Vol. 3, (Fried R., ed). pp 1–73. Marcel Dekker, Inc., New York.

Baxter C. F. and Roberts E. (1959) Elevation of γ-aminobutyric acid in rat brain with hydroxylamine. *Proc. Soc. Exp. Biol.* **101**, 811–815.

Bengtsson G., Idham G., and Westerdahl G. (1981) Glass capillary gas chromatographic analyses of free amino acids in biological microenvironments using electron capture or selected ion-monitoring detection. *Anal. Biochem.* **111**, 163–175.

Berl S. and Frigyesi T. L. (1969) The turnover of glutamate, glutamine, aspartate and GABA-labeled with $[1 - {}^{14}C]$acetate in caudate nucleus, thalamus and motor cortex (cat). *Brain Res.* **12**, 444–455.

Bertilsson L. and Costa E. (1976) Mass fragmentographic quantitation of glutamic acid and γ-aminobutyric acid in cerebellar nuclei and sympathetic ganglia of rats. *J. Chromatogr.* **118**, 395–402.

Bertilsson L., Mao C. C., and Costa E. (1977). Application of principles of steady-state kinetics to the estimation of γ-aminobutyric acid turnover rate in nuclei of rat brain. *J. Pharmacol. Exp. Ther.* **200**, 277–284.

Blass J. P. (1960) The simple monosubstituted guanidines of mammalian brain. *Biochem. J.* **77**, 484–489.

Bloom F. E. and Iversen L. L. (1971) Localizing ^{3}H-GABA in nerve terminals or rat cerebral cortex by electron microscopic autoradiography. *Nature,* London **229**, 268–630.

Cattabeni F., Galli C. L., De Angelis L., and Maggi A. (1977) Mass fragmentography as a tool for enzyme activity assay: Measurement of glutamate decarboxylase, in Quantitative Mass Spectrometry in Life Sciences, (DeLeenheer A. P., and Rancucci R. R., eds.), pp. 237–244.Elsevier, Amsterdam.

Cattabeni F., Galli C. L., and Eros T. (1976) A simple and highly sensitive mass fragmentographic procedure for γ-aminobutyric acid determinations. *Anal. Biochem.* **72**, 1–7.

Cheney D. L. (1984) *Drug Effects on Transmitter Dynamics: An overview, in*

Dynamics of Neurotransmitter Function, (Hanin I., ed.), pp 63–80. Raven, New York.

Colby B. N. and McCamen M. W. (1978) A highly specific and sensitive determination of gamma-aminobutyric acid by gas chromatography-mass spectrometry. *Biomed. Mass Spect.* **5**, 215–219.

Colby B. N. and McCamen M. W. (1979) A comparison of calculation procedures for isotope dilution determinations using gas chromatography-mass spectrometry. *Biomed. Mass Spect.* **6**, 225–230.

Corradetti R., Moneti G., Moroni F., Pepeu G., and Wieraszko A. (1983) Electrical stimulation of the striatum radiation increases the release and neosynthesis of aspartate, glutamate, and γ-aminobutyric acid in rat hippocampal slices. *J. Neurochem.* **41**, 1518–1525.

Costa E., Guidotti A., Moroni F., and Peralta E. (1979) Glutamic acid as a transmitter precursor and as a transmitter. in, *Glutamic Acid: Advances in Biochemistry and Physiology,* (Filer L. J. *et al.,* eds.), pp 151–161, Raven, New York.

Coutts R. T., Jones G. R., and Liu, S-F. (1979) Quantitative gas chromatography/mass spectrometry of trace amounts of glutamic acid in water samples. *J. Chromatogr. Sci.* **17**, 551–554.

Dessort D., Maitre M., and Mandel P. (1982) γ-Aminobutyric acid (GABA) turnover in neurons and glial cell cultures of chicken embryo. *C. R. Acad. Sci.,* (Paris) **295**, 537–546.

Falkner F. C. (1981) Comments on some common aspects of quantitative mass spectrometry. *Biomed. Mass Spect.* **8**, 43–46.

Faull K. F. and Barchas J. D. (1983) Negative-ion mass spectrometry fused-silica capillary gas chromatography of neurotransmitters and related compounds. *Methods Biochem. Anal.* **29**, 325–383.

Faull K. F. and DoAmaral J. R., Berger D. A. and Barchas J. D. (1978) Mass spectrometric identification and selected ion monitoring quantitation of γ-aminobutyric acid (GABA) in human lumbar cerebrospinal fluid. *J. Neurochem.* **31**, 1119–1122.

Ferkany J. W., Smith L. A., Seifert W. E. Jr., Caprioli R. M., and Enna S. J. (1978) Measurement of gamma-aminobutyric acid (GABA) in blood. *Life Sci.* **22**, 2121–2128.

Finlayson P. J., Christopher R. K., and Duffield A. M. (1980) Quantitation of fourteen urinary α-amino acids using isobutane gas chromatography chemical ionization mass spectrometry with ^{13}C amino acids as internal standards. *Biomed. Mass Spect.* **7**, 450–453.

Fonnum F. (1981) The turnover of transmitter amino acids, with special reference to GABA, in Central Amino Acid Transmitter Turnover, (Pycock C. J. and Tabener P. V., eds.), pp 105–124. University Park Press, Baltimore.

Freeman M. S., Co C., Mote T. R., Lane J. D., and Smith J. E. (1980) Determination of content and specific activity of amino acids in central nervous system tissue utilizing tritium and carbon-14 dual labelling. *Anal. Biochem.* **106**, 191–194.

Gaffney T. E., Hammar C-G., Holmstedt B., and McMahon, R. E. (1971) Ion specific detection of internal standards labelled with stable isotopes. *Anal. Chem.* **43**, 307–310.

Guidotti A., Cheney D. L., Trabucchi M., Doteuchi M., Wang C., and Hawkins R. A. (1974) Focussed microwave radiation: A technique to minimize post mortem changes of cyclic nucleotides, dopa and choline and to preserve brain morphology. *Neuropharmacology* **13**, 1115–1122.

Halliday D. and Rennie M. J. (1982) The use of stable isotopes for diagnosis and clinical research. *Clin. Science* **63**, 485–496.

Haseyawa Y., Ono T., and Maruyama Y. (1981) γ-Aminobutyric acid in the murine brain: Mass fragmentographic assay method and postmortem changes. *Jpn. J. Pharmacol.* **31**, 165–173.

Hassler R., Haug P., Nitsch C., Kim J. S., and Paik K. (1982) Effect of motor and premotor cortex ablation on concentrations of amino acids, monoamines, and acetylcholine and on the ultrastructure in rat striatum. A confirmation of glutamate as the specific corticostriatal transmitter. *J. Neurochem.* **38**, 1087–1098.

Haymond M. W., Howard C. P., Miles J. M., and Gerish J. E. (1980) Determination of leucine flux in vivo by gas chromatography-mass spectrometry utilizing stable isotopes for trace and internal standard. *J. Chromatogr.* **183**, 403–409.

Holdiness M. R., Justice J. B., Salamone J. D., and Neill D. B. (1981) Gas chromatographic-mass spectrometric determination of glutamic acid decarboxylase activity in subregions of rat brain. *J. Chromatogr.* **225**, 283–290.

Huizinga J. D., Teelken A. W., Muskiet F. H. J., Jeuring H. J., and Wolthier B. G. (1978) Gamma-aminobutyric acid determination on human cerebrospinal fluid by mass-fragmentography. *J. Neurochem.* **30**, 911–913.

Irving C. S., Nissim I., and Lapidot A. (1978) The determination of amino acid pool sizes and turnover rates by gas chromatographic mass spectrometric analyses of stable isotope enrichment. *Biomed. Mass Spect.* **5**, 117–122.

Iwase H., Takeuchi Y., and Murai A. (1979) Gas chromatography-mass spectrometry of trimethylsilyl derivatives of amino acids. *Chem. Pharm. Bull.* **27**, 1307–1315.

Johnson J. L. (1972) Glutamic acid as a synaptic transmitter in the nervous system, A review. *Brain Res.* **37**, 1—19.

Kingston E. E. and Duffield A. M. (1978) Plasma amino acid quantitation using gas chromatography chemical ionization mass spectrometry and ^{13}C amino acids as internal standards. *Biomed. Mass Spect.* **5**, 621–626.

Krnjevic L. (1974) Chemical nature of synaptic transmission in vertebrates. *Physiol. Rev.* **54**, 418–540.

Lane J. D., Sands M. P., Freeman M. E., Cherek D. R., and Smith J. E.

(1982) Amino acid neurotransmitter utilization in discrete rat brain regions is correlated with conditioned emotional response. *Pharmacol. Biochem. Behav.* **16**, 329–340.

Lapidot A. and Nissim I. (1980) Regulation of pool sizes and turnover rates of amino acids in humans: ^{15}N-glycine and ^{15}N-alanine single-dose experiments using gas chromatography-mass spectrometry analysis. *Metabolism* **29**, 230–239.

Lapin A. and Karobath M. (1980) Simple and rapid mass fragmentographic method for the determination of glycine in brain tissue. *J. Chromatogr.* **193**, 95–99.

Lovell R. A., Elliott S. J., and Elliott K. A. C. (1963) The γ-aminobutyric acid and factor 1 content of brain. *J. Neurochem.* **10**, 479–488.

Mackenzie S. L. and Tenaschuk D. (1979a) Quantitative formation of N(O,S)-heptafluorobutyryl isobutyl amino acids for gas chromatographic analyses. I. Esterification, *J. Chromatogr.* **171**, 195–208.

Mackenzie S. L. and Tenaschuk D. (1979b) Quantitative formation of N(O,S)-heptafluorobutyryl isobutyl amino acids for gas chromatographic analyses. II. Acylation, *J. Chromatogr.* **173**, 53–63.

Matthews D. E. and Hayes J. M. (1976) Systematic errors in gas chromatography-mass spectrometry isotope ratio measurements. *Anal. Chem.* **48**, 1375–1382.

Matthews D. E., Conway J. M., Young V. R., and Bier D. M. (1981a) Glycine nitrogen metabolism in man. *Metabolism* **30**, 886–893.

Matthews D. E., Bier D. M., Rennie M. J., Edwards R. H. T., Halliday D., Millward D. J., and Clugston G. A. (1981b) Regulation of leucine metabolism in man: A stable isotope study. *Science* **214**, 1129–1131.

Mao C. C., Peralta E., Moroni F., and Costa E. (1978) The turnover rate of γ-aminobutyric acid in the substantia nigra following electrical stimulation or lesioning of the strionigral pathways. *Brain Res.* **155**, 147–152.

Mee J. M. L., Korth J., and Halpern B. (1977) Rapid and quantitative blood amino acid analyses by chemical ionization mass spectrometry. *Biomed. Mass Spect.* **4**, 178–181.

Minard F. N. and Mushahwar I. K. (1966) Synthesis of γ-aminobutyric acid from a pool or glutamic acid in brain after decapitation. *Life Sci.* **5**, 1409–1413.

Moroni F., Lombardi G., Moneti G., and Cortesini C. (1983) The release and neosyntheses of glutamic acid are increased in experimental models of hepatic encephalopathy. *J. Neurochem.* **40**, 850–854.

Moroni F., Tanganelli S., Bianchi C., Moneti G., and Beani L. (1980) A mass-fragmentographic approach to release studies of endogenous GABA, glutamic acid and glutamine *in vitro*. *Pharmacol. Res. Comm.* **12**, 501–505.

Moroni F., Bianchi C., Tanganelli S., Moneti G., and Beani L. (1981) The release of γ-aminobutyric acid, glutamate and acetylcholine from striatal slices: A mass fragmentographic study. *J. Neurochem.* **36**, 1691–1697.

Murayama K., Shindo N., Mineki R., and Ohta K. (1981) Determination of glutamic acid and γ-aminobutyric acid in Ringer's solution without desalination at the femtomole level by gas chromatography chemical ionization mass spectrometry. *Biomed. Mass Spect.* **8**, 165–169.

Okada Y., Nitsch-Hassler C., Kim J. S., Bak I. J., and Hassler R. (1971) Role of γ-aminobutyric acid (GABA) in the extrapyramidal motor system. I. Regional distribution of GABA in rabbit, rat, guinea pig, and baboon CNS. *Exp. Brain Res.* **13**, 514–518.

Perry T. L., Hansen S., and Gandham S. S. (1981) Postmortem changes of amino compounds in human and rat brain. *J. Neurochem.* **36**, 406–412.

Petty F., Tucker H. N., Molinary S. V., Flynn N. W., and Wander J. D. (1976) Quantitation of glycine in plasma and urine by chemical ionization mass fragmentography. *Clin. Chim. Acta.* **66**, 111–117.

Quastel J. H. (1974) Amino acids and the brain. *Biochem. Soc. Trans.* **2**, 765–780.

Racagni G., Cheney D. L., Trabucchi M., Wang C., and Costa G. (1974) Measurement of acetylcholine turnover rate in discrete areas of the rat brain. *Life Sci.* **15**, 1961–1975.

Revuelta A., Cheney D. L., and Costa E. (1981) Measurement of γ-aminobutyric acid turnover rates in brain nuclei as an index of interactions between γ-aminobutyric acid and other transmitters, in Glutamate as a Neurotransmitter, (Dichiara G. and Gessa G. L., eds.), pp 169–181. Raven Press, New York.

Richardson, D. L. and Scudder, C. L. (1976) Microwave irradiation and brain γ-aminobutyric acid levels in mice. *Life Sci.* **18**, 1431–1440.

Robitaille Y., Wood P. L., Etienne P., Lal S., Finlayson M. H., Gauthier S., and Nair N. P. V. (1982) Reduced cortical choline acetyl transferase activity in Gerstmann-Straussler syndrome. *Prog. NeuroPsychopharmacol. & Biol. Psychiatr.* **6**, 529–531.

Sander G., Dessort D., Schmidt P., and Karli P. (1981) Distribution of GABA in the periaqueductal gray matter. Effects of medial hypothalamus lesions. *Brain Res.* **224**, 279–280.

Schulman M. F. and Abramson F. P. (1975) Plasma amino acid analyses by isotope ratio gas chromatography mass spectrometry computer techniques. *Biomed. Mass Spect.* **2**, 9–14.

Shank R. P. and Aprison M. H. (1971) Postmortem changes in the content and specific radioactivity of several amino acids in four areas of the rat brain. *J. Neurobiol.* **2**, 145–151.

Shaw R. K. and Heine J. D. (1965) Ninhydrin positive substances present in different areas of normal rat brain. *J. Neurochem.* **12**, 151–155.

Sjoquist B. (1979) Analysis of tyrosine and deuterium-labelled tyrosine in tissues and body fluids. *Biomed. Mass Spect.* **6**, 392–395.

Spencer H. J., Tominez G., and Halpern B. (1981) Mass spectrographic analysis of stimulated release of endogenous amino acids from rat hippocampal slices. *Brain Res.* **212**, 194–197.

Summons R. E., Pereira W. E., Reynolds W. E., Rindfleisch T. C., and Duffield A. M. (1974) Analyses of twelve amino acids in biological fluids by mass fragmentograhy. *Anal. Chem.* **46,** 582–585.

Van Den Berg C. J., Matheson D. F., Ronda G., Reijnierse G. L. A., Blokhuis G. G. D., Kroon M. C., Clarke D. D., and Garfinkel D. (1975) A model of glutamate metabolism in brain: A biochemical analysis of a heterogenous structure, in Metabolic Compartmentation and Neurotransmission, (Berl S., Clarke D. D., and Schneider O., eds.), pp 515–558. Plenum Press, New York.

Wolfensberger M. and Amsler U. (1982) Mass fragmentographic method for the determination of trace amounts of putative amino acid neurotransmitters and related compounds from brain perfusates collected *in vivo. J. Neurochem.* **38,** 451–456.

Wolfensberger M., Redweik U., and Curtius H. Ch. (1979) Gas chromatography-mass spectrometry and selected ion monitoring of the N,N′-dipentafluoropropionylhexafluoroisopropyl ester of glutamine. *J. Chromatogr.* **172,** 471–475.

Wolfensberger M., Reubi J. C., Canzek V., Redweik U., Curtius H. Ch., and Cuenod M. (1981) Mass fragmentographic determination of endogenous glycine and glutamic acid released *in vivo* from the pigeon optic tectum. Effect of electric stimulation of a midbrain nucleus. *Brain Res.* **224,** 327–336.

Wood P. L. (1982) A selected ion monitoring assay for dopamine and its metabolites using negative chemical ionization. *Biomed. Mass Spect.* **9,** 302–306.

Zaylak M. -J., Curtius H-Ch, Leimbacher W., and Redweik U. (1977) Quantitation of deuterated and non-deuterated phenylalanine and tyrosine in human plasma using the selective ion monitoring method with combined gas chromatography-mass spectrometry. Application to the *in vivo* measurement of phenylalanime-4-monooxygenase activity. *J. Chromatogr.* **142,** 523–531.

Zilversmit D. B. (1960) The design and analysis of isotope experiments. *Am. J. Med.* **29,** 832–848.

Chapter 4

Double-Isotope Dansyl Microassay for Cerebral Amino Acids

ROGER F. BUTTERWORTH

1. Introduction

1-Dimethylamino-naphthalene-5-sulfonyl chloride (dansyl chloride, Fig. 1) was originally used by Weber in 1952 for the formation of fluorescent conjugates of albumin. Dansyl chloride reacts with primary and secondary amino groups and with phenols.

Dansyl derivatives exhibit an intense yellow fluorescence and can be detected at extremely low concentrations, the sensitivity being comparable to that encountered using radioactive isotopes. In favorable circumstances, concentrations as low as 0.005 nM of an amine can be determined quantitatively by direct fluorimetry on thin-layer plates (Seiler, 1970).

1.1. Dansylation of Amino Acids

Dansylation of most amino acids (Fig. 2) is optimal in the pH range 9.5–11.0. pH values higher than 11.0 are associated with increased hydrolysis, and at values below 8.0, the unreacted (protonated) form of the amino group predominates and reaction rates are slow (Seiler, 1970). The extent of dansylation of an amino acid depends not only on the pH of the reaction mixture, but also on the chemical structure of the amino acid. Steric hindrance around the amino group, as is present in certain branched-chain amino acids, leads to decreased stability of the dansyl derivative. In addition, rates of dansylation depend on the excess quantity of dansyl chloride used, as well as on the reaction time and temperature. Since dansyl chloride is only slightly soluble in water,

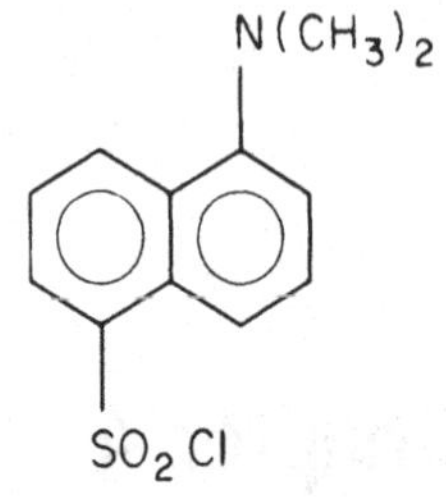

1- dimethylamino - naphthalene - 5 - sulfonyl chloride

(dansyl chloride)

Fig. 1.   Structure of dansyl chloride.

dansylation reactions are generally performed in acetone–water mixtures.

Dansyl chloride is a rather nonselective reagent and is suitable, therefore, for use in the quantitative estimation of amino acids in biological samples only when appropriate methods of separation are applied. To date, methods used to separate dansyl amino acids have included paper chromatography, paper electrophoresis, thin-layer chromatography (TLC), and, more recently, high-performance liquid chromatography (HPLC). Methods of separation of dansyl amino acids have been the subject of a review article (Seiler, 1970). Boulton and Bush (1964) successfully separated the dansyl amino acids from a peptide hydrolysate on paper chromatograms. Seiler and Wiechmann (1964) introduced TLC as an alternative method of separation that has been successfully used by numerous laboratories throughout the world. Separated dansyl amino acids are detected on thin-layer chromatograms, under UV irradiation, as intense yellow fluorescent spots.

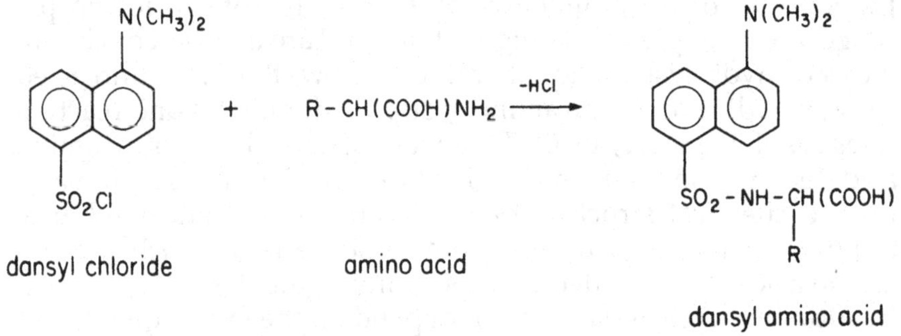

Fig. 2.   Derivatization of amino acids with dansyl chloride.

In recent years, HPLC has increasingly found an application as an efficient means of separation of dansyl amino acids (*see*, for example, Chapman et al., 1982).

However, although it is possible to detect very low concentrations of dansyl amino acids fluorimetrically, use of the dansyl reaction for the quantitative analysis of amino acids has proven difficult in the past. Many factors influence the yield of the dansylation reaction. In addition, side reactions are known to occur: During dansylation, a second molecule of dansyl chloride may combine with the acidic group of the amino acid and the product then readily hydrolyzes in the alkaline conditions of the reaction medium (Boulton, 1968). This situation is potentially even more complex when mixtures of amino acids are to be derivatized. In such cases, the yield of the dansylation reaction for each amino acid constituent of the mixture must be estimated. This has required, in the past, time-consuming parallel experiments involving dansylation of standard amino acid mixtures.

Such difficulties are overcome by the use of [^{3}H]-labeled dansyl chloride and the inclusion of ^{14}C-labeled amino acids as internal standards.

1.2. The Double-Isotope Dansyl Assay

The principle of the double-isotope dansyl assay is as follows: A sample containing amino acids in unknown quantities is added to a mixture of ^{14}C-labeled amino acids, and the resulting mixture is dansylated using [^{3}H]-dansyl chloride. Following separation of the dansylated amino acids, generally by thin-layer chromatography, the ratio dpm [^{3}H]:dpm (^{14}C) for each dansyl amino acid is measured. This ratio bears a linear relation to the amount of unlabeled amino acid present in the original sample (*see* section 2.4).

2. The Double-Isotope Dansyl Microassay for Measurement of Cerebral Amino Acids

2.1. Appropriate Sacrifice-Dissection Conditions

Many recent studies have addressed the question of postmortem stability of cerebral amino acids. Techniques for appropriate sacrifice of the experimental animal and brain dissection depend on the amino acid to be measured and the brain structure under investigation. One recent study reported the effects of increasing death-to-freezing intervals on amino acid content of rat brain (Perry et al ., 1981). It was found that brain concentrations of taurine, glutamine, glutamate, and aspartate did not vary appreciably between brain immersed in liquid nitrogen within 30 s of death and brain allowed to stand for 24 h at 4°C. Glycine remained stable for only 30 min at 4°C and GABA increased rapidly following death. This latter finding confirms several previous reports. In 1956, Elliott and Florey reported that GABA content of brain tissue increased by 30–40% 2 min after excision. Similar increases have been observed by others (Shank and Aprison, 1971; Alderman and Shellenberger, 1974).

Two procedures have been described that minimize postmortem increases of cerebral GABA, namely, rapid freezing of the brain *in situ* [using either liquid nitrogen or the freeze-blowing technique (Veech et al., 1973)], or focused microwave irradiation (Guidotti et al., 1974). There are advantages and disadvantages of these sacrifice-dissection techniques. On the one hand, the distinctive visibility of certain nuclei (e.g., substantia nigra) may be lost in microwaved brain tissue (Iadarola et al., 1979). On the other hand, liquid nitrogen freezing may not be adequate to rapidly fix GABA levels in deep regions of the brain. It has been estimated that it takes 50–90 s to cool deep regions of the brain to 0°C after immersion in liquid nitrogen (Ferrendelli et al., 1972). The freeze-blowing sacrifice technique is applicable only when measurements of whole brain are to be performed.

In addition to the problem of postmortem stability of cerebral GABA, other possible sources of variability have been reported. These include interspecies differences (Alderman and Shellenberger, 1974) and circadian variations of GABA in certain brain structures (Cattabeni et al., 1978). In view of these sources of potential variability, it is important in experiments in which regional GABA concentrations are to be measured to control for the following parameters:

(a) age, sex, and strain of rats
(b) time of day at sacrifice
(c) sacrifice-dissection conditions

Dissection times should be as short as possible and should be rigorously matched in control and experimental groups. For regions requiring longer dissection times, animals should be sacrificed by microwave irradiation, or, alternatively, control experiments of postmortem GABA changes as a function of time in that particular brain structure should be performed.

2.2. Extraction and Dansylation of Amino Acids

The sample of nervous tissue under investigation is homogenized in cold perchloric acid and, following centrifugation, the supernatant is used for dansylation. It is reportedly advantageous in certain circumstances to separate amino acids from tissue homogenates prior to dansylation. This approach is useful in following the specific activity of amino acids formed from labeled precursors such as ^{14}C-glucose and acetate, and permits an estimation of the radioactivity present in "the total amino acid fraction" prior to dansylation (Beart and Snodgrass, 1975). The homogenate is applied to a column of cation exchange resin (Dowex 50 or Amberlite H$^+$, for example). The amino acid fraction is subsequently eluted using triethylamine ($1M$ in 50% methanol) and evaporated to dryness.

Following the addition of appropriate amounts of ^{14}C-labeled amino acids, the pH of the mixture is adjusted to pH 9–9.5. This is generally accomplished using either potassium carbonate or triethylamine. Use of the latter reportedly leads to a reduction of the quantity of salt transferred to the chromatogram, resulting in improved separation of certain amino acids (Brown and Perham, 1973). [^{3}H]-Dansyl chloride is then added to the mixture of ^{14}C-amino acids and supernatant containing endogenous amino acids and dansylation is allowed to proceed in the dark, generally at room temperature, for 30 min.

2.3. Thin-Layer Chromatographic Separation of Dansyl Amino Acids

Separation of the mixture of dansyl amino acids is generally accomplished using two-dimensional thin-layer chromatography. Chromatograms are developed in the first dimension in formic acid (1.5% v/v). In the second dimension, the solvent system generally used is benzene:acetic acid (9:1 v/v). Alternatively, *n*-butyl acetate:methanol:acetic acid (20:3:1 v/v) has been employed (Beart and Snodgrass, 1975). Both of these systems have been used to successfully separate dansyl glycine, GABA, glutamate, aspartate, and glutamine (Beart and Snodgrass, 1975; Joseph and Halliday, 1975). Undansylated amino acids migrate with the solvent front in the first dimension. In order to adequately resolve dansyl taurine, plates are subjected to a single development in the second dimension using 35% ammonia. This results in effective separation of dansyl taurine from dansyl hydroxide (Joseph and Halliday, 1975).

Separated dansyl amino acids are visualized under UV light. "Spots" are marked and removed. Elution of the dansyl amino acid from the chromatogram is then accomplished using a strong organic base such as triethylamine or hyamine hydroxide. Hyamine hydroxide is reportedly better than solvents or solvent mixtures such as *n*-butyl acetate:methanol, which do not effectively elute dansyl glutamate and dansyl aspartate (Joseph and Halliday, 1975).

2.4. Calculation of Results Using Double Isotope Dansyl Microassay

As expected from theoretical considerations, a linear relationship exists between the ratio dpm [^{3}H]:dpm (^{14}C) and the amount of unlabeled amino acid added (Brown and Perham, 1973; Joseph and Halliday, 1975). This relationship is derived as follows:

If *a* nmol of ^{14}C-amino acid having specific activity A is mixed with *x* nmol of unlabeled amino acid, the specific activity of the resulting mixture will fall to:

$$ A \left(\frac{a}{a + x} \right) $$

If this mixture is now allowed to react with ^{3}H-dansyl chloride to yield a monodansylated derivative of specific activity *B*, the ratio dpm [^{3}H]:dpm (^{14}C) in the product will be:

$$ \frac{B}{A \left(\dfrac{a}{a + x} \right)} $$

That is,

$$ \text{dpm } [^3\text{H}]: \text{dpm } (^{14}\text{C}) = \frac{B}{A \left(\dfrac{a}{a + x} \right)} $$

$$ \frac{B}{A} + \frac{B}{A} \left(\frac{x}{a} \right) $$

$$ = \frac{B}{Aa} x + \frac{B}{A} $$

Compare with:

$$ y = mx + c $$

Thus, a linear calibration curve may be obtained by adding various known amounts (x) of unlabeled amino acid to constant amounts of ^{14}C-labeled amino acid and measuring the ratio dpm [^{3}H]:dpm (^{14}C) in the dansylated product. The ratio dpm [^{3}H]:dpm (^{14}C) will depend, for a given dansyl amino acid, only on the specific activity of the [^{3}H]-dansyl chloride used for derivatization and on the specific activity of the ^{14}C-amino acid reacting with it. Thus, once the ^{14}C-amino acid internal standard has been mixed with the extract containing the endogenous amino acid, the ratio dpm [^{3}H]/dpm (^{14}C) will not be affected by subsequent manipulations, such as the amount of material applied to the chromatogram, or by factors such as the yield of the dansylation reaction. In addition, any ambiguous areas on the chromatogram (for example, where one spot is adjacent to a second one) may be safely omitted (Joseph and Halliday, 1975).

2.5. Example Protocol Using the Double-Isotope Dansyl Assay

The method described here for the simultaneous estimation of glutamate, aspartate, GABA, glutamine, glycine, and taurine in regions of the rat brain is essentially that originally described by Joseph and Halliday (1975), with minor modifications (Brown and Perham, 1973; Butterworth et al., 1982).

2.5.1. Materials

Chemicals, solvents. All chemicals and solvents used for the procedures described are analytical grade, unless otherwise stated.

Water. Double glass-distilled water is used throughout for the preparation of standard solutions and buffers.

10% Hyamine hydroxide. A 225.1-mL volume of methyl benzethonium hydroxide (hyamine hydroxide) (1.0M) in 1 L methanol.

Toluene/PPO/POPOP. A 12.75-g quantity of 2,5-diphenyloxazole (PPO), plus 0.336 g 1,4-bis-2-2-(5-phenyloxazalyl)-benzene (POPOP) in 3 L toluene (scintillation grade).

Radiolabeled compounds (Purchased from New England Nuclear, Inc.):

L-Glutamic acid (^{14}C-U), specific activity > 225 mCi/ mmol

L-Glutamine (^{14}C-U), specific activity > 200 mCi/mmol

GABA (^{14}C-U), specific activity > 170 mCi/mmol
L-Glycine (^{14}C-U), specific activity > 90 mCi/mmol
Taurine (^{14}C), specific activity > 90 mCi/mmol
Dimethylamino-1-naphthalene sulfonyl chloride 5-(methyl ^{3}H), specific activity > 10 Ci/mmol

2.5.2. Sacrifice-Dissection Procedure

Following sacrifice by decapitation into liquid nitrogen, heads are allowed to warm up to $-20°$C in the cabinet of a freezing micro-tome. Brains are removed and dissected between -10 and $-20°$C, according to the guidelines of Glowinski and Iversen (1966), to af-ford material from the following seven major brain structures: cerebral cortex, hippocampus, striatum, hypothalamus, mid-brain, cerebellum, and medulla-pons. When strictly adhered to, these sacrifice-dissection conditions are adequate to minimize postmortem amino acid changes. Using these conditions, re-gional content of cerebral amino acids (including GABA) are found to be in very good agreement with values obtained in ani-mals sacrificed using focused microwave irradiation (*see*, for ex-ample, Butterworth et al., 1982, Chapman et al., 1982). Tissue is stored at $-80°$C until time of assay.

2.5.3. The Double-Isotope Dansyl Microassay

Weighed fragments of nervous tissue are separately homoge-nized in 10 vol $HClO_4$ (0.48M) at $4°$C. The homogenate is centri-fuged (20,000g, 15 min, $4°$C), and the supernatant removed and stored on ice. A blank (0.48M $HClO_4$) and standard solutions of amino acids are carried through in parallel with tissue extracts. A summary of the method is shown schematically in Fig. 3.

Tissue extract (10 µL) is transferred to a 0.3-mL "ReactiVial" (Pierce Chemical Co., USA). To this extract is added 2 µL of a mix-ture of ^{14}C-labeled amino acids containing the following:

40 nCi GABA; specific activity, 205 nCi/nmol
40 nCi ^{14}C-glycine; specific activity, 109 nCi/nmol
57 nCi ^{14}C-aspartate; specific activity, 203 nCi/nmol
40 nCi ^{14}C-glutamate; specific activity, 296 nCi/nmol
23 nCi ^{14}C-glutamine; specific activity, 254 nCi/nmol
50 nCi ^{14}C-taurine; specific activity, 104 nCi/nmol

To the carefully mixed solution is added 2 µL K_2CO_3 (2M) and the precipitate removed by centrifugation (3000g, 10 min, $4°$C).

To each tissue extract, blank, or standard solution containing ^{14}C-internal standards, prepared as described above, is added 10

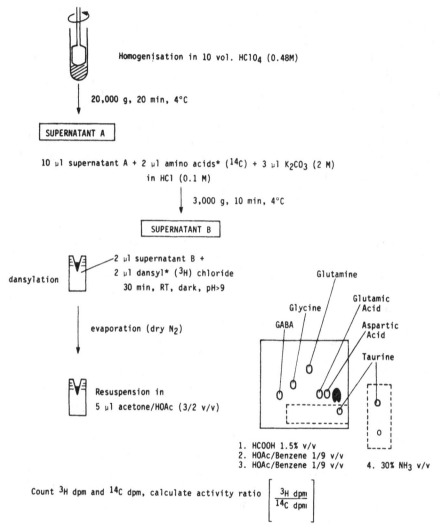

Fig. 3. Schematic representation of double-isotope dansyl microassay for measurement of amino acids in brain.

μL [³H]-Dansyl chloride. [³H]-dansyl chloride of the required specific activity is prepared by mixing [³H]-dansyl chloride (250 μCi/mL; specific activity, 18.9 μCi/nmol) in acetone at 4°C with cold dansyl chloride in acetone (14.8 mM) (Pierce Chemical Co., USA) in the ratio 4:1 v/v. The final concentration of diluted [³H]-dansyl chloride is then 29.76 nmol/10 μL, specific activity, 66.95 nCi/nmol. This solution (10 μL) is evaporated to dryness under dry N_2 and the residue is resuspended in 2 μL dry acetone at

4°C. To this solution of [^{3}H]-dansyl chloride is added 2 μL of the tissue extract, blank, or standard solution. Dansylation is allowed to proceed at room temperature for 30 min in the dark. Solvents are then removed by evaporation (dry N$_2$) and the residue is resuspended in 5 μL acetone:acetic acid (3:2 v/v) at 4°C.

Thin-layer chromatography is performed on 5 × 5 cm micropolyamide sheets (Schleicher and Schuell, USA). Approximately 0.5 μL of the mixture of dansyl amino acids is applied, using a 1-μL Hamilton syringe, to the lower left-hand corner of the chromatogram. After air drying, chromatograms are developed in closed chromatography tanks that have been previously equilibrated. Chromatography in the first dimension is performed using formic acid (1.5% v/v). After careful drying, chromatography in the second dimension is performed using acetic acid:benzene (9:1 v/v). After drying, the chromatogram is developed a second time (and third time, if necessary) in order to adequately resolve dansyl aspartate and dansyl glutamate. The separation is monitored under UV light after each migration. The chromatogram is then cut (*see* Flow Sheet, Fig. 3) and dansyl taurine separated by development in the second dimension using 35% ammonia (v/v).

Dansyl amino acid "spots" are identified by circling with a pencil. They are then cut into liquid scintillation vials, each containing 1 mL hyamine hydroxide (1M in methanol). Care is taken, where there is overlapping of spots, not to include any areas of ambiguity. Elution of dansyl amino acids is allowed to proceed overnight, after which 10 mL of a scintillation cocktail containing PPO/POPOP/toluene is added. Using a dual channel liquid scintillation counter, cpm [^{3}H] and cpm (^{14}C) are obtained for each sample. Counting efficiencies are determined by the external reference method and dpm [^{3}H] and dpm (^{14}C) are computed. Tissue concentrations of amino acids are then determined from calibration curves using the standard amino acid solutions.

2.6. Neurochemical Applications of the Double-Isotope Dansyl Microassay

The double isotope dansyl assay has been used extensively to measure amino acids in nervous tissue. In Table 1, glutamate concentrations in regions of the adult rat CNS are compared using the technique described in section 2.5.3. Data is taken from the published study of Butterworth et al., 1979.

Glutamate levels were highest in cerebral cortex and hippocampus, and lowest in retina and caudally-situated regions of the

TABLE 1
Glutamate Content of Rat CNS Regions Using the Double
Isotope Dansyl Microassay[a]

CNS region	Glutamate concentration, µg/g wet wt ± SEM ($n = 12$)
Olfactory bulbs	6.91 ± 0.35
Cerebral cortex	11.76 ± 0.38
Hippocampus	10.97 ± 0.25
Striatum	8.81 ± 0.37
Midbrain	8.27 ± 0.37
Hypothalamus	7.10 ± 0.29
Cerebellum	9.86 ± 0.20
Medulla-pons	5.03 ± 0.50
Spinal cord (cervical)	4.13 ± 0.08
Spinal cord (dorsal)	3.44 ± 0.47
Spinal cord (lumbo-sacral)	4.22 ± 0.26
Retina	4.24 ± 0.38

[a]From Butterworth et al., 1979.

CNS (medulla, pons, spinal cord). The assay has been applied to the measurement of amino acids in segments of autopsied human spinal cord; patients dying with the clinical diagnosis of Friedreich's ataxia were found to have selectively decreased concentration of glutamate and glutamine in gray matter of lumbar spinal cord (Butterworth and Giguère, 1984).

Free amino acids in pigeon optic nerve have been accurately measured using the double-isotope dansyl microassay (Beart and Snodgrass, 1975). Only 50 µg of tissue was required for the analysis. Improved separation of dansyl glycine, alanine, and GABA was achieved using the solvent system *n*-butyl acetate:-methanol:acetic acid (20:3:1 v/v) for development of the chromatogram in the second dimension. Endogenous amino acids in segments of the rat retina have also been measured using the assay (Morjaria and Voaden, 1979).

The double-isotope dansyl microassay has also been used to measure release of amino acids from brain preparations in vitro (Snodgrass and Iversen, 1973). Rat cortical tissue was chopped using a McIlwain chopper and suspended in Krebs–bi-carbonate–glucose medium. Electrical stimulation was applied and, following superfusion, the medium was changed for a Krebs-bicarbonate buffer containing 55 m*M* KCl. In this way, efflux of amino acids during electrical stimulation or potassium depolarization was studied.

Comparison of data obtained using the double-isotope microassay with that using other techniques is summarized in Table 2.

2.7. Advantages and Disadvantages of the Double-Isotope Dansyl Assay

Advantages of the double-isotope dansyl assay may be summarized as follows:

(a) Errors caused by incomplete derivatization are eliminated.

(b) Problems caused by loss of dansyl amino acid resulting from manipulation of the sample and incomplete extraction into the scintillation fluid are also eliminated.

(c) The highly fluorescent nature of dansyl amino acid derivatives permits ease of visualization, thereby precluding the necessity for autoradiographic scanning of chromatograms.

(d) The technique is very sensitive; it has been employed for the estimation of cerebral amino acids in microgram quantities of tissue and to measure release of

TABLE 2

Amino Acid Content of Rat Cerebellum Using Double-Isotope Dansyl Microassay: Comparison with Other Methods

Amino acid	Amino Acid Concentration, μmol/g wet wt $\pm$ SEM		
	Amino acid analyzer[a] ($n = 10$)	Thin-layer chromatography of dinitrophenyl derivatives[b] ($n = 12$)	Double-isotope dansyl microassay[c] ($n = 7$)
Aspartate	2.19 $\pm$ 0.18	1.99 $\pm$ 0.06	2.78 $\pm$ 0.17
Glutamate	8.42 $\pm$ 0.75	9.68 $\pm$ 0.16	9.71 $\pm$ 0.19
Glutamine	2.55 $\pm$ 0.58	4.74 $\pm$ 0.15	5.89 $\pm$ 0.13
GABA	1.56 $\pm$ 0.37	1.58 $\pm$ 0.07	2.36 $\pm$ 0.14
Glycine	1.04 $\pm$ 0.12	0.63 $\pm$ 0.03	0.89 $\pm$ 0.03
Taurine	4.89 $\pm$ 0.46	6.18 $\pm$ 0.04	4.90 $\pm$ 0.42

[a]From Shaw and Heine (1965).
[b]From Shank and Aprison (1970).
[c]From Hamel et al. (1979).

endogenous amino acids from brain slices in vitro.

(e) The cost/analysis is substantially less than that using a conventional amino acid analyzer.

(f) The technique requires no sophisticated equipment other than a liquid scintillation counter.

The major disadvantage of the double-isotope dansyl microassay is its requirement for considerable manual manipulation.

3. Summary

The double-isotope dansyl microassay represents a double-labeling technique for the measurement of amino acids in small quantities of nervous tissue. The method is based on the formation of [^{3}H]-dansyl derivatives of cerebral amino acids using [^{3}H]-dansyl chloride. ^{14}C-labeled amino acids are added as internal standards to circumvent problems caused by incomplete derivatization. Amino acids in microgram quantities of nervous tissue can be accurately measured by this method. An example protocol is described in which the amino acids glutamate, aspartate, glutamine, GABA, glycine, and taurine may be measured in regions of the rat brain. The method has been employed in the past for the measurement of amino acids in optic nerve and segments of retina, as well as in dissected regions of spinal cord. The technique is also applicable to studies of release of endogenous amino acids from brain preparations in vitro, as well as for the study of amino acid compositions of peptides.

In addition to its high sensitivity, the method is reliable, relatively inexpensive, and does not require the use of sophisticated equipment other than a liquid scintillation counter. The method does, however, require considerable manual manipulation.

Values for amino acid content of rat brain regions using the double-isotope dansyl microassay are comparable to those obtained using other techniques, such as the amino acid autoanalyzer.

Acknowledgments

The author is grateful to The Medical Research Council of Canada and Fonds de Recherche en Santé du Québec for continuing support. The author thanks Françoise Trotier for her assistance with the preparation of the manuscript.

References

Alderman J. L. and Shellenberger M. K. (1974) γ-Aminobutyric acid (GABA) in the rat brain: re-evaluation of sampling procedures and the postmortem increase. *J. Neurochem.* **22**, 937–940.

Beart P. M. and Snodgrass S. R. (1975) The use of a sensitive double-isotope dansylation technique for amino acid analysis. *J. Neurochem* **24**, 821–824.

Boulton A. A. and Bush I. E. (1964) Qualitative and quantitative analysis of amino acids and their dansyl derivatives. *Biochem. J.* **92**, 11P.

Boulton A. A. (1968) The Automated Analysis of Absorbent and Fluorescent Substances Separated on Paper Strips, in *Methods of Biochemical Analysis*, Vol. 16 (Glick D., ed.) Interscience, New York, pp. 327–340.

Brown J. P. and Perham R. N. (1973) A highly sensitive method for amino-acid analysis by a double-isotope-labeling technique using dansyl chloride. *Eur. J. Biochem* **39**, 69–73.

Butterworth R. F., Hamel E., Landreville F., and Barbeau, A. (1979) Amino acid changes in thiamine-deficient encephalopathy: some implications for the pathogenesis of Friedreich's ataxia. *Can. J. Neurol. Sci.* **6**, 217–222.

Butterworth R. F., Merkel A. D., and Landreville F. (1982) Regional amino acid distribution in relation to function in insulin hypoglycemia. *J. Neurochem.* **38**, 1483–1489.

Butterworth R. F. and Giguère J. F. (1984) Amino acids in autopsied human spinal cord: selective changes in Friedreich's ataxia. *Neurochem. Pathol.* **2**, 7–18.

Cattabeni F., Maggi A., Monduzzi M., De Angelis L., and Racagni G. (1978) GABA: circadian fluctuations in rat hypothalamus. *J. Neurochem.* **31**, 565–567.

Chapman A. G., Riley K., Evans M. C. and Meldrum B. S. (1982) Acute effects of sodium valproate and γ-vinyl GABA on regional amino acid metabolism in the rat brain: incorporation of $2\text{-}^{14}C$-glucose into amino acids. *Neurochem. Res.* **7**, 1089–1105.

Elliott K. A. C. and Florey E. (1956) Factor 1-inhibitory factor from brain. Assay conditions in brain. Stimulating and antagonizing substances. *J. Neurochem.* **1**, 181–191.

Ferrendelli J. A., Gay M. H., Sedgwick W. G., and Chang M. M. (1972) Quick freezing of the murine CNS: Comparison of regional cooling rates and metabolite levels when using liquid nitrogen or Freon-12. *J. Neurochem.* **19**, 979–987.

Glowinski J. and Iversen L. L. (1966) Regional studies of catecholamines in the rat brain. I. The disposition of [^{3}H]-norepinephrine, [^{3}H]-dopamine, and [^{3}H]-dopa in various regions of the brain. *J. Neurochem.* **13**, 655–669.

Guidotti A., Cheney D. L., Trabucchi M., Doteuchi M., Wang C. T., and Hawkins R. A. (1974) Focused microwave radiation: a technique to minimize postmortem changes of cyclic nucleotides, dopa, and choline and to preserve brain morphology. *Neuropharmacol.* **13**, 1115–1122.

Hamel E., Butterworth R. F., and Barbeau A. (1979) Effect of thiamine deficiency on levels of putative amino acid transmitters in affected regions of the rat brain. *J. Neurochem.* **33**, 575–577.

Iodarola M. J., Raines A., and Gale K. (1979) Differential effects of *n*-dipropylacetate and amino-oxyacetic acid on γ-aminobutyric acid levels in discrete areas of rat brain. *J. Neurochem.* **33**, 1119–1123.

Joseph M. H. and Halliday J. (1975) A dansylation microassay for some amino acids in brain. *Anal. Biochem.* **64**, 389–402.

Morjaria B. and Voaden J. M. (1979) The formation of glutamate, aspartate, and GABA in the rat retina; glucose and glutamine as precursors. *J. Neurochem.* **33**, 541–551.

Perry T. L., Hansen S., and Gandham S. S. (1981) Postmortem changes of amino compounds in human and rat brain. *J. Neurochem.* **36**, 406–412.

Seiler N. and Wiechmann J. (1964) Zum nachweis von aminosauren im 10^{-10} molmap stab. Trennung von 1-dimethyl-aminonaphthalin-5-sulfonyl aminosauren auf dunnschicht-chomatogrammen. *Experientia* **20**, 559.

Seiler N. (1970) Use of the Dansyl Reaction in Biochemical Analysis, in *Methods of Biochemical Analysis,* Vol. 18, Interscience, New York, pp. 259–337.

Shank R. P. and Aprison M. H. (1970) The metabolism in vivo of glycine and serine in eight areas of the rat cerebral nervous system. *J. Neurochem.* **17**, 1461–1475.

Shank R. P. and Aprison M. H. (1971) Postmortem change in the content and specific radioactivity of several amino acids in four areas of the rat brain. *J. Neurobiol.* **2**, 145–151.

Shaw R. K. and Heine J. D. (1965) Ninhydrin positive substances present in different areas of normal rat brain. *J. Neurochem.* **12**, 151–155.

Snodgrass S. R. and Iversen L. L. (1973) A sensitive double-isotope derivative assay to measure release of amino acids from brain in vitro. *Nature (New Biol.),* **241**, 154–156.

Veech R. L., Harris R. L., Veloso D., and Veech E. H. (1973) Freeze-blowing: a new technique for the study of brain in vivo. *J. Neurochem.* **20**, 183–188.

Weber G. (1952) Polarization of the fluorescence of macromolecules. *Biochem. J.* **51**, 155–167.

Chapter 5

Liquid Chromatographic Determination of Amino Acids After Precolumn Fluorescence Derivatization

PETER LINDROTH, ANDERS HAMBERGER, AND
MATS SANDBERG

1. Introduction

Certain amino acids are now well established as the transmitters employed by the majority of neurons in the central nervous system (Curtis and Johnston, 1974; Watkins and Evans, 1981; Fagg and Foster, 1983). Investigation of the physiology, pharmacology, and biochemistry of these amino acids in relation to neurotransmission demands their quantification at low concentrations in both tissue extracts and extracellular fluids of the brain. Amino acids in physiological fluids are conventionally determined by ion-exchange chromatography and postcolumn derivatization (Stein and Moore, 1954). The resolution is well characterized and the technique is sensitive when fluorogenic reagents are used. However, an analysis time up to several hours is generally required for separation of amino acids. Over the last 10 yr, liquid chromatography, previously named high-performance liquid chromatography (Ettre, 1981), has experienced explosive growth. The use of precolumn fluorogenic derivatization in combination with reversed phase liquid chromatography has simplified and improved the separation of amino acids. This system today constitutes an attractive alternative to the conventional amino acid analyzers (Lindroth and Mopper, 1979; Ejnarsson et al., 1983).

The system provides high sensitivity and eliminates the requirement for cleanup steps in sample handling. Furthermore, the analysis of amino acids in physiological fluids can be performed in less than 30 min (Lindroth and Mopper, 1979; Hamberger et al., 1981).

2. Derivatization Reagents

The number of reagents for derivatization of amino compounds is larger than for any other class of compounds. Two of the most useful reagents for precolumn fluorescence labeling in liquid chromatography are dansyl chloride and o-phthaldialdehyde (OPA).

2.1. Dansyl Chloride

Dansyl chloride was, for a long time, the most widely used fluorogenic derivatization reagent for amino acid determination. However, there are certain problems connected with the dansyl reaction at low amino acid concentrations (Neadle and Pollitt, 1965). Since dansyl chloride and its byproducts are fluorescent, serious interference can occur at high sensitivity levels.

2.2. o-Phthaldialdehyde

The use of OPA for the determination of primary amino acids was originally reported by Roth in 1971. In the presence of a reducing agent, such as 2-mercaptoethanol, OPA reacts in alkaline medium with primary amines, to form highly fluorescent thioalkyl-substituted isoindoles (Simons and Johnson, 1976, Fig. 1). OPA as a derivatization reagent has several advantages:

- It is water-soluble
- It is nonfluorescent *per se*
- The fluorescent products have a high quantum yield that results in high sensitivity
- The fluorescent products are rapidly formed at room temperature

The main disadvantages are:

- Only primary amines react to give fluorophores
- A relative instability of some derivatives

The OPA reaction, however, may be used to detect the secondary amines hydroxyproline and proline if they are oxidized

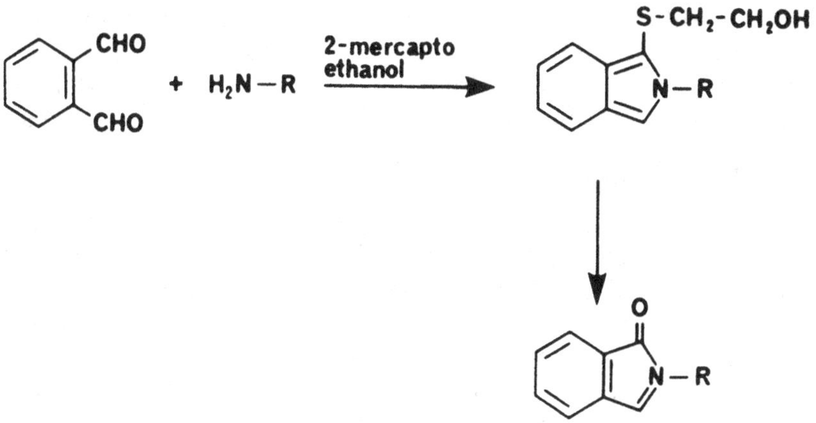

Fig. 1. Reaction of a primary amine with OPA and 2-mercaptoethanol. The chemical conversion to nonfluorescent derivatives (bottom) is shown by the vertical arrow.

prior to derivatization (Lee and Drescher, 1978). The instability of the derivatives is the result of an intermolecular rearrangement (Simons and Johnson, 1976). In order to obtain high reproducibility despite the instability of some fluorescent derivatives, a fixed reaction time of 1 or 2 min is employed (Lindroth and Mopper, 1979). Simons and Johnson (1977) showed that the stability of the OPA derivatives increased when ethanethiol was used instead of 2-mercaptoethanol. Furthermore, the fluorescent products were stable in 95% ethanol for 4 d. Hill et al. (1979) found that the derivatization products are stable for a least 1 h with a high methanol content in the reaction mixture. As physiological samples in most cases are diluted before the reaction with OPA, methanol would be a suitable solvent. In this case, only a few microliters can then be injected without occasional band broadening at the top of the column.

The poor fluorescence of the cysteine, lysine, and ornithine derivatives may be a drawback of the technique. Cysteine yields weakly fluorescent properties due to its sulfydryl group (Cooper and Turnell, 1982). However, these sulfydryl groups can be blocked with iodoacetic acid, iodoacetamine, or acrylonitrile, with the result that fluorescent isoindoles can then be formed with OPA (Cooper and Turnell, 1982). Cysteine may also be oxidized to cysteic acid, which forms a highly fluorescent product with OPA. Following the oxidation of cysteine, however, it may be difficult to obtain high reproducibility with the OPA derivatization of amino acids, because the conditions for these two reactions are different (Cooper and Turnell, 1982). Addition of a surfactant to the OPA reagent can be used to increase the fluorescence response (Gardner and Miller, 1980) and the stability (Jones et al., 1981) of the lysine and the ornithine derivatives.

2.3. Other Derivatives

Since OPA does not yield a fluorescent product with secondary amines, other derivatizing reagents have been employed for this purpose. Recently, reagents based on nitrobenzofurazan have been described, among which 7-chloro-4-nitrobenzofurazan (NBD chloride) is commonly used because of its faster reaction with secondary amines than with primary amines (Roth, 1978; Ahnoff et al., 1981; Umagat et al., 1982). NBD chloride is stable, nonfluorescent and readily soluble in organic solvents. It reacts slowly with amines in alkaline solutions even at high temperatures (50–75°C). The reaction time, however, can be reduced to 5 min if NBD fluoride is used (Imai and Watanabe, 1981). Moreover, water solubility of the reagent can be increased with NBD ethers (Johnson et al., 1982).

Both primary and secondary amino acids react with 9-fluorenylmethyl chloroformate (FMOC chloride) under alkaline, aqueous conditions (Anson-Moye and Boning, 1979). The reaction is complete in 30 s. However, although FMOC chloride is fluorescent, the excess reagent and the byproducts are easily removed by extraction prior to separation (Ejnarsson et al., 1983).

3. Automation of Precolumn Derivatization

The instabilitiy of some OPA derivatives rules out the use of a traditional autoinjector, the derivatives needing to be formed immediately prior to injection. Venema et al. (1983) have described a continuous flow system in which the sample is mixed with OPA reagent and led through the loop of the injection valve under the control of a microprocessor. Recently, Waters Associates introduced an automated precolumn derivatization system. Using a WISP Sample Processor, equipped with an additional valve, the OPA reagent and the sample are withdrawn into a sample loop and automatically applied to the column without any interference with the normal flow. In our laboratory, the reproducibility of this system was ±6.2 and ±5.4% for glutamine and alanine , respectively ($n = 11$, SD), with a dialysate collected in vivo (*see* below). The dead volume is about 100 µL, including a mixing column filled with glass beads that, as the only negative effect, has a noncritical band broadening of aspartate. Even though the dialysate mentioned above was stable for an overnight run, other extracts or physiological fluids must be checked for stability.

4. Separation of Neuroactive Amino Acids—Methodological Considerations

4.1. Liquid Chromatographic System

Reversed phase liquid chromatography on C-18 bonded phase has often been used for separation of amino acid derivatives. As these columns can be different from one supplier to another, it is impossible to recommend the conditions for the mobile phase. Lindroth and Mopper (1979) and Hill et al. (1979) used phosphate buffer at different pH values and methanol in the mobile phase.

In the report on the separation of a mixture of OPA-derivatized amino acids, phosphate buffers in the pH range 6.3–7.7 were used (Lindroth and Mopper, 1979). Two amino acids of importance in neurotransmission, glutamine and GABA, coelute with histidine and alanine, respectively, in this system. Complete separation of the amino acids was obtained, however, at pH 5.25.

A spectrum of 34 OPA-derivatized primary amines is shown in Fig. 2. The total elution time is less than 30 min and the intersample time is 40 min. The elution is accomplished by gradually increasing the methanol content in the elution solvent. It should be noted, however, that some compounds are coeluted in the routine analysis set-up, e.g., glycine comigrates with arginine, threonine, and glycero-phosphoethanolamine. They may, however, be separated with other elution paradigms (Jones and Gilligan, 1983). In addition, tetrahydrofuran added to the mobile phase improves the separation of glycine and threonine (Hodgin, 1979).

Coelution problems were also encountered with β-alanine (Sandberg and Corazzi, 1983), which migrates with homocarnosine and hypotaurine at pH 5.25. Its detection in extracts was also hampered by taurine, which overshadowed the β-alanine peak. The problem is solved by lowering the pH to 3.5, which (a) separates β-alanine from homocarosine and hypotaurine, and (b) retains β-alanine relative to taurine, thus ensuring a better quantification of β-alanine.

The separation of asparagine and glutamate in the standard mixture is taken as an index of how the pH of the buffer should be

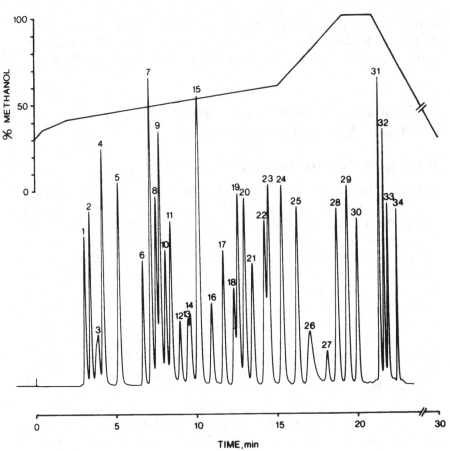

Fig. 2. The elution pattern of a standard mixture of OPA-derivatized primary amines, separated on a 5 μm Nucleosil C-18 column (200 × 4.6 mm id). The flow-rate was 1 mL/min employing the indicated gradient of methanol and Na phosphate buffer (50 mM, pH 5.25). Each peak represents 39 pmol except for those indicated below. 1, glutathione; 2, cysteic acid; 3, O-phosphoserine (19.5 pmol); 4, cysteine sulfinic acid; 5, aspartic acid; 6, asparagine (19.5 pmol); 7, glutamic acid; 8, histidine; 9, serine; 10, glutamine; 11, 3-methyl-histidine; 12, α-aminoadipic acid (9.8 pmol); 13, citrulline (9.8 pmol); 14, carnosine; 15, threonine,glycine; 16, O-phosphoethanolamine; 17, taurine (19.5 pmol); 18, β-alanine (19.5 pmol); 19, tyrosine; 20, alanine; 21, α-aminoisobutyric acid; 22, β-aminoisobutyric acid; 23, γ-amino-n-butyric acid; 24, β-amino-n-butyric acid; 25, α-amino-butyric acid; 26, histamine; 27, cystathione (19.5 pmol); 28, methionine; 29, valine; 30, phenylalanine; 31, isoleucine; 32, leucine; 33, 5-hydroxytryptamine (5-HT); 34, lysine. The chromatographic system consisted of a Varian LC 5000 chromatograph and a Schoeffel FS 970 fluorimeter.

set for a particular batch of column support. If glutamate is eluted too close to asparagine, the pH is lowered in small steps (i.e., with more H^+, glutamate is eluted further back in the spectrum). The elution may be accelerated by a steeper increase of methanol concentration in the gradient.

4.2. Tissue Extracts and Perfusates

For routine extraction, CNS tissue, frozen in liquid nitrogen is homogenized in ice-cold $0.6M$ perchloroacetic acid (PCA). After centrifugation of the protein precipitate ($10,000g$ for 10 min), the extracts are either used directly or neutralized with K_2HCO_3 and stored frozen (after an additional centrifugation). The presence of trichloroacetic acid (TCA) in a sample distorts the separation and TCA is preferably extracted with diethyl ether (3 ×). Sulfosalicylic acid (SSA) gives a peak in the beginning of the chromatogram, but does not interfere with the separation. Homogenization in PCA is the best procedure in our hands, because it does not interfere at all with the OPA system. Before derivatization of the amino acids, the extracts are normally diluted with H_2O to contain 1–10 μmol/L of amino acid. Levels of free amino acids in the cerebral cortex, the cerebellum, and the brain stem were analyzed with an ion-exchange automatic analyzer and ninhydrin detection in parallel with liquid chromatography system and these levels are shown in Fig. 3. The correlation between liquid and ion exchange chromatography was highly satisfactory.

Depending on the perfusion system, varying amounts of proteases and other enzymes may be present in the perfusates and these amounts determine the handling of the samples. The dialysates and perfusates are normally collected in small plastic tubes, frozen, and stored at −20°C for a limited period of time.

4.3. Contamination

One problem which is particularly critical when small quantities of amino acids are measured is the risk of contamination with extraneous amino acids (aspartate, glutamate, serine, glycine, and alanine) from the glassware, solutions, and so forth (Lee and Drescher, 1978). Obvious precautions necessary to follow include the use of disposable gloves, freshly distilled water, and highly pure chemicals. Amino acids in the blanks are further reduced if the glassware and plastic collection tubes are prerinsed several times with diluted phosphoric acid and distilled water. In order to avoid contamination from the injection loop, syringe, and elution buffer, the reaction mixture may be acidified after a set time (Jones et al., 1981).

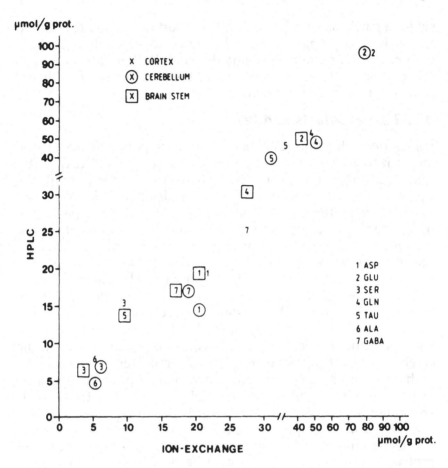

Fig. 3.   Correlative analysis of free amino acids in extracts from cerebral cortex, cerebellum, and brain stem with the OPA liquid chromatography and ion-exchange techniques.

4.4. Internal Standards

As a check for correct derivatization and pipeting, it is advisable to include an internal standard with the sample (preferably as early as possible in the sample treatment), e.g., β-amino-*n*-butyric acid, which is eluted in the middle of the chromatogram. Compounds that are eluted in other regions of the spectra, e.g., cysteate or cystein sulfinate, may also be used. For routine quantification we compare peak heights with known amounts of external and internal standards.

The separation and shape of some peaks are distorted after consecutive sample loading, and this may be due either to increased dead volume at the top of the column and/or to accumulation of noneluted material. Therefore, we retop the column daily and "wash" with pure methanol for at least 2 min between each sample. This gives good reproducibility and a column lasts for at least one thousand separations. A replaceable guard column may also be used to save the analytical column (Jones et al., 1981).

4.5. Identification

Mass spectrometry is certainly required for correct identification of an unknown peak (Larson, 1982). Less complicated ways may, however, be employed to test the homogeneity of a peak: (1) Elution with different gradient systems and at different pHs can also be used. (2) Quantification should be similar with different derivatives, e.g., those formed with OPA and FMOC chloride (Ejnarsson et al., 1983), and with different principles of chromatography, i.e. reversed-phase and ion-exchange; (3) An attractive method was recently described by Joseph and Davies (1983), with fluorimetric and electrochemical detection in series, confirmation of the identities of the OPA derivatives was obtained in a single run.

4.6. Radioactive Amino Acids

With the postcolumn derivatization technique, a split of the effluent before derivatization enables easy measurement of the radioactive underivatized amino acid. The precolumn derivatization technique with OPA may be equally simple, but some aspects need to be considered and examined.

The instability and chemical conversion of some OPA derivatives imply that a derivatized compound may, in fact, result in one fluorescent and two radioactive peaks (Simson and Johnson, 1976, Fig. 1). The chemical rearrangement of the derivatives may, however, be a minor factor with respect to retention and the fluorescent and nonfluorescent derivatives may coelute. The use of more chemically stable amino acid derivatives, i.e. those formed by reaction with FMOC chloride, eliminates this problem. When the radioactivity of an amino acid is measured, it is often desirable and necessary to inject larger concentrations of amino acids than in a routine experiment. With the OPA method it is then critical to: (a) make sure that OPA is present in the required molar excess (Lindroth and Mopper, 1979), (b) lower the pH of the reagent mixture to spare the top of the column, and (c) use the same or lower proportion of organic solvent in the sample as in the beginning of the gradient in order to obtain a concentration of the derivatives on the column top.

5. Neurochemical Applications

The potential applications of the precolumn derivatization techniques in neurochemistry are, of course, considerable. A few examples concerning release of amino acids in vitro and in vivo and levels of amino acids in CNS fluids will be reported. Finally, the usefulness of the method in the clinical routine, i.e. for determination of amino acids in cerebrospinal fluid, will be touched upon briefly.

In earlier studies, the efflux of exogenously applied, radioactive amino acids was most commonly studied (Katz et al., 1969; Srinivasan et al., 1969). The high sensitivity of the postcolumn derivatization technique with fluorigenic derivatization made it possible to follow the release of a range of endogenous amino acids during various conditions (Norris et al., 1980). Moreover, the precolumn derivatization techniques (Lindroth and Mopper, 1979; Ejnarsson et al., 1983) enable measurement of subpicomol amounts of amino acids.

5.1. Amino Acids in the Visual System

In order to characterize the transmitters in the optic nerve, we have monitored the release of amino acids from tissue slices of the superior colliculus (SC). The levels of aspartate, glutamate, GABA, taurine, and β-alanine in the perfusate increased during depolarization with either 56 mM potassium (Fig. 4) or veratridine. This increased release was Ca-dependent (Note: not for taurine) and tetr odotoxin-sensitive (Sandberg and Jacobson, 1981).

Mainly because of its neuroactive properties, β-alanine was previously proposed to be a neurotranmitter (DeFeudis and Martin del Rio, 1977). However, certain data suggested a glial localization of β-alanine in some parts of the CNS (Schon and Kelly, 1975), whereas other reports indicated both glial and neuronal uptake of radioactive β-alanine (Brunn et al., 1974). The demonstration of release of endogenous β-alanine suggests a neuronal localization of β-alanine in the SC and a role in visual neurotransmission. This is compatible with reports on regional differences in β-alanine concentration (Martin del Rio et al., 1977) and the apparent regional specificity for release of exogenously applied β-alanine (O'Fallon et al., 1981). Our results illustrate the necessity of a sensitive detection system for studying the release of putative transmitter substances. The chemical identity of β-alanine from the SC has been checked by using the system at different pHs and by ion-exchange chromatography. As β-alanine is a constituent of the dipeptides carnosine and anserine (Sano, 1970),

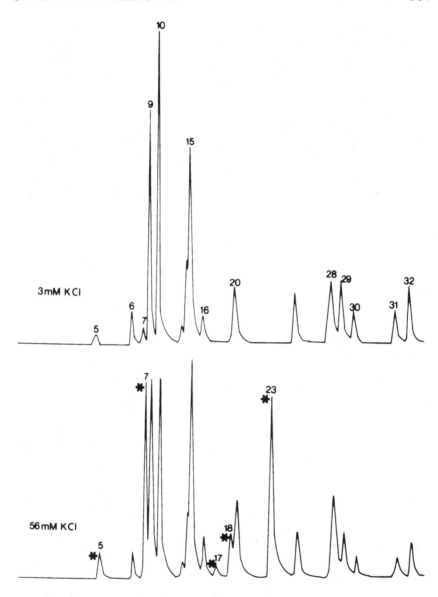

Fig. 4. The spectrum of primary amines collected by superfusion of superior colliculus slices in vitro during nonstimulatory conditions (3 m*M* KCl) and during chemical depolarization of the tissue (56 m*M* KCl). The following substances were released in increased amounts: 5, aspartic acid; 7, glutamic acid; 17, taurine; 18, β-alanine; and 23, γ-aminobutyric acid (GABA). The elution gradient used was *not* identical to that decribed in Fig. 2 that resulted in comigration of tyrosine and alanine. The β-alanine peak, therefore, is eluted closer to alanine. Modified from Sandberg et al. (1982).

one might hypothesize that the observed release of β-alanine could be secondary to extracellular proteolytic activity following efflux of one or both of these dipeptides. However, no concomitant increase in the other constituents of the dipeptides (L-1-methylhistidine and histidine) was observed in our studies on the visual system.

In parallel, dialysates were collected in vivo from the lateral geniculate nucleus of the cat with a closed dialysis sampling probe (*see* below) (Sandberg and Lindstrom, 1983). An attached electrode was used for recording and stimulation. Local electrical stimulation evoked a pronounced efflux of aspartate, glutamate, GABA, and taurine (Fig. 5).

In order to localize the origin of the release, we performed perfusion and depolarization experiments on SC tissue slices after degeneration of the optic nerve terminals (Sandberg and Corazzi, 1983). The Ca-dependent release of neuroactive amino acids was, however, unchanged, which implies that the major transmitter of the optic nerve is not an easily detectable primary amine.

5.2. Brain Dialysis—Amino Acids in the Extracellular Fluid of the Living Animal

The combination of in vivo perfusion and analysis of endogenous compounds secreted by CNS cells into the extracellular compartment represents a research area in which the rapid and sensitive OPA technique is extremely valuable. The development of probes suitable for withdrawing minute fluid samples from defined CNS regions largely without interfering with the normal activity of the animal actually began 30 years ago. Such techniques were of limited value, previously, as the amounts recovered were below the detection levels of, e.g., the conventional amino acid analyzers. The best known of the in vivo "probes" is probably the push–pull cannula (Gaddum, 1961), but other approaches have also been used, such as the cortical cup (Dodd et al., 1974) and dialysis systems (Bito et al., 1966; Delgado et al., 1972; Ungerstedt et al., 1982; Hamberger et al., 1982). With the OPA system, transmitter amino acids and other amine compounds can be monitored virtually on-line.

As a closed perfusion system may be precalibrated to give actual levels of free amino acids, the composition in the extracellular fluid may be compared with that in the cerebrospinal fluid (CSF; Hamberger et al., 1983; 1985). A comparison between the hippocampus extracellular fluid and the cisternal fluid in albino rabbits showed good agreement with respect to taurine, valine, leucine, isoleucine, and phenylalanine, whereas the concentrations of ex-

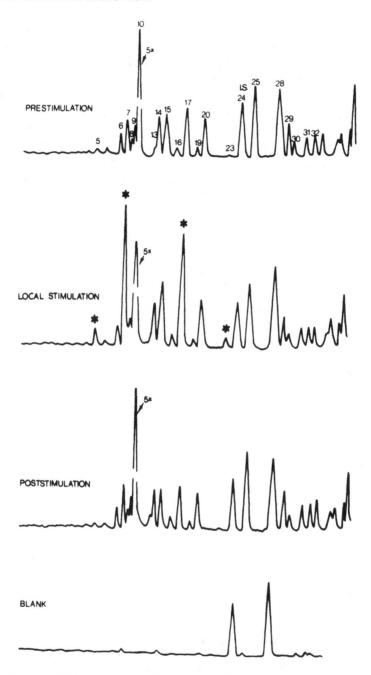

Fig. 5. The spectrum of primary amines collected by brain dialysis of the dorsolateral geniculate nucleus of the cat in vivo during nonstimulated conditions. These were collected both before and after the local stimulation that was delivered through an electrode attached to the dialysis sampling probe. The elution gradient, time-scale, and other

tracellular glutamate, glutamine, alanine, asparagine, and ornithine were 50% of the CSF levels (Table 1). The interplay between
the surfaces of the CNS, the plexus choroideus, and the drainage
systems from which the composition of the CSF is finally derived
remains unresolved, but there is an increasing awareness that the
microenvironment of the brain cells is different from CSF.

The in vivo probes have given an insight into the excitation-
coupled changes in extracellular amino acids, findings which differ considerably from those in tissue slice perfusion experiments.
Probably because of more intact uptake processes in vivo, the visible liberation of neuroactive compounds is decreased, as is also
the loss of poorly accumulated amino acid compounds that occurs
in in vitro experiments. The quantitatively most prominent shifts

TABLE 1

Free Amino Acids in Plasma, Hippocampal Extracellular Fluid,
and Cerebrospinal Fluid (CSF)[a]

Amino acid	Plasma ($n = 7$)	Extracell. fluid ($n = 8$)	CSF ($n = 6$)
Tau	48 ± 6	6 ± 0.5	6 ± 1.4
PEA[a]	24 ± 3	6 ± 0.7	12 ± 0.9
Ser	275 ± 25	36 ± 2.8	116 ± 11
Asn	60 ± 6	3 ± 0.2	16 ± 1.6
Glu	56 ± 9	3 ± 0.8	10 ± 2.2
Gln	821 ± 56	226 ± 26.9	583 ± 27
Ala	361 ± 41	14 ± 1.4	39 ± 5.2
Val	179 ± 19	11 ± 1.3	14 ± 1.2
Ileu	112 ± 12	8 ± 0.9	7 ± 0.7
Leu	91 ± 13	8 ± 1.0	7 ± 0.3
Phe	46 ± 5	5 ± 0.5	7 ± 0.6

[a]Results (mean and SE) are expressed in μmol/L.
[b]PEA = phosphoethanolamine

Fig. 5 (cont.) chromatographic parameters are identical to those described in Fig. 2. Efflux of the following substances (denoted by an aster-
isk) was substantially increased during local stimulation: 5, aspartic acid;
7, glutamic acid; 17, taurine; and 23, GABA. A blank collected before the
insertion of the dialysis probe into the brain is also shown. Modified
from Sandberg and Lindstrom (1983). Note that α-amino-n-butyric acid
(25) is coeluted with ethanolamine and that methionine (28) is coeluted
with ammonia using this elution paradigm.

in amino acid concentrations that occur during neuroexcitation in vivo concern glutamine and taurine, whereas glutamate, aspartate, and GABA change strikingly in tissue-slice experiments. The latter group is kept under tight control in vivo, but is easily mobilized with specific uptake inhibitors (Hamberger et al., 1983).

5.3. Amino Acids in Cerebrospinal Fluid

The conventional amino acid analysis with colorimetric detection often resulted in conflicting data on the normal amino acid profile of the CSF and hence on the use of CSF amino acid levels in diagnostics (Hare et al., 1980). The introduction of fluorimetric, radioreceptor, gas chromatographic, and mass-spectrometric methods, alone or in combination, has improved the analytical strength of many routine laboratories. We recently utilized the OPA technique for the analysis of CSF amino acids in an unselected group of patients (n = 32) from the Department of Neurology of the university hospital (Table 2). Various aspects of

TABLE 2
Concentrations of Amino Acids (μmol/L) in
Cerebrospinal Fluid[a]

Amino acid	Mean	Maximum	Minimum
Asp	7	13	4
Glu	0.5	2	0
Gln	595	765	450
Ser	41	58	18
Gly + Arg	34	60	21
PEA	5	9	2
Tau	8	19	4
Tyr	10	18	5
Ala	49	76	31
Met	5	10	3
Val	21	41	13
Phe	11	17	8
Ileu	10	17	5
Leu	13	21	8
Lys	30	46	21

[a]Summary of concentrations of amino acid in unselected patient material (n = 32). Means, maximal and minimal concentrations are shown. Nonacidified samples stored at $-20°C$ were used. PEA = phosphoethanolamine.

fluid handling been have studied, including concentration gradients and the acid deproteinization step (Lundquist et al., in preparation). The reproducibility of the technique for these purposes was highly satisfactory. It is particularly notable that free glutamate levels are markedly increased after deproteinization. The large-scale capacity of the system when fully automated adds a considerable potential to clinical neurological diagnosis.

6. Conclusions

Precolumn derivatization techniques represent a large step towards simple, ultrasensitive amino acid analysis. This chapter is written during a period of rapid development in this technology. Columns and detection systems are already available offering faster and more sensitive analysis. Theoretically, a few amino acid molecules approaching the postsynaptic neuron can already be measured, provided they can be delivered to the detection system. The neurochemist is faced with the challenge to develop fine enough probes to ensure that the amino acid data are part of a structure–function relationship. Consequently, an urgent need for developing neurochemistry is to approach the degree of resolution already obtained in neuroscience disciplines related to ultrastructure and electrophysiology.

Acknowledgments

We gratefully acknowledge the secretarial assistance of Ms. Gull Grönstedt and Ms. Margareta Dahl. This work was supported by grants from the Swedish Medical Research Council (12X-00164), Axelson Johnsons Foundation and the Swedish Natural Science Research Council (1905-100).

References

Ahnoff M., Grundevik I., Arfwidsson A., Fonselius J., and Persson B.-A. (1981) Derivatization with 4-chloro-7-nitrobenzofurazan for liquid chromatographic determination of hydroxyproline in collagen hydrolysate. *Anal. Chem.* **53,** 485–489.
Anson Moye H. and Boning A. J. (1979) A versatile fluorogenic labelling reagent for primary and secondary amines: 9-fluorenylmethyl chloroformate. *Anal. Lett.* **12,** 25–35.

Bito L., Davson H., Levin E., Murray M., and Snider N. (1966) The concentrations of free amino acids and other electrolytes in cerebrospinal fluid, in vivo dialysate of brain and blood plasma of the dog. *J. Neurochem.* **13,** 1057–1067.

Bruun A., Ehinger B., and Forsberg A. (1974) In vitro uptake of β-alanine into rabbit retinal neurons. *Exp. Brain Res.* **19,** 239–247.

Cooper J. D. H. and Turnell D. C. (1982) The fluorescence detection of cystine by *o*-phthaldialdehyde derivatization and its separation using high performance liquid chromatography. *J. Chromatogr.* **227,** 150–161.

Curtis D. R. and Johnston G. A. R. (1974) Amino acid transmitters in the mammalian central nervous system. *Ergebn. Physiol.* **69,** 94–188.

DeFeudis F. V. and Martin del Rio R. (1977) Is β-alanine an inhibitory neurotransmitter? *Gen. Pharmacol.* **8,** 177–180.

Delgado J. M. R., DeFeudis F. V., Roth R. H., Ryngo D. K., and Mitruka B. M. (1972) Dialytrode for long-term intracerebral perfusion in awake monkeys. *Arch Int. Pharmacodyn.* **198,** 9–21.

Dodd P. R., Pritchard M. J., Adams R. C. F., Bradford, H. F., Hicks G., and Blanchard K. C. (1974) A method for the continuous, long-term superfusion of the cerebral cortex of unanaesthetized, unrestrained rats. *J. Physiol.* (Lond.) **7,** 897–907.

Ejnarsson S., Josefsson B., and Lagerkvist S. (1983) Determination of amino acids using 9-fluorenylmethyl chloroformate and reversed phase HPLC. *J. Chromatogr.* **282,** 609–618.

Ettre L. S. (1981) The nomenclature of chromatography II. Liquid chromatography. *J.Chromatogr.* **220,** 29–63.

Fagg G. E. and Foster A. C. (1983) Amino acid neurotransmitters and their pathways in the mammalian central nervous system. *Neurosci.* **9,** 701–719.

Gaddum J. H. (1961) Push–pull cannulae. *J. Physiol.* (Lond.) **155,** 1–2.

Gardner W. S. and Miller W. H. (1980) Reverse-phase liquid chromatographic analysis of amino acids after reaction with *o*-phthaldialdehyde. *Anal. Biochem.* **101,** 61–65.

Hamberger A., Jacobson I., Lindroth P., Mopper K., Nyström B., Sandberg M., Molin S.-O., and Svanberg U. (1981) Neuron-glia interactions in the biosynthesis and release of transmitter amino acids. *Adv. Biochem. Psychopharmacol.* **29,** 509–518.

Hamberger A., Jacobson I., Molin S.-O., Nyström B., Sandberg M., and Ungerstedt U. (1982) Metabolic and transmitter compartments for glutamate, in: *Neurotransmitter Interaction and Compartmentation* (Bradford H. F. ed.), pp. 359–378. Plenum Press, New York.

Hamberger A., Berthold C.-H., Karlsson B., Lehmann A., and Nyström B. (1983) Extracellular GABA, glutamate and glutamine in vivo - perfusion-dialysis of the rabbit hippocampus. *Neurology and Neurobiology* **7,** 473–492.

Hamberger A., Berthold C.-H., Jacobson I., Karlsson B., Lehmann A., Nyström B., and Sandberg M. (1985) In vivo brain dialysis of extracellular non-transmitter and putative transmitter amino acids, in: *In Vivo Release of Neuroactive Substances in the CNS: Methods, Findings and Perspectives* (Bayon A. and Drucker Colin R., eds.). Academic Press, New York.

Hare, T. A., Manyam N. V. B., and Glaeser B. S. (1980) Evaluation of cerebrospinal fluid γ-aminobutyric acid content in neurological and psychiatric disorders, in: *Neurobiology of Cerebrospinal Fluid*, Vol. 1 (Wood, J., ed.) pp. 171–187. Plenum Press, New York and London.

Hill D. W., Walters F. H., Wilson T. D., and Stuart J. D. (1979) High-performance liquid chromatographic determination of amino acids in the picomole range. *Anal. Chem.*, **51**, 1338–1341.

Hodgin J. C. (1979) The separation of pre-column *o*-phthaldialdehyde derivatized amino acids by high-performance liquid chromatography. *J. Liq. Chromatogr.* **2**, 1047–1059.

Imai K. and Watanabe Y. (1981) Fluorimetric determination of secondary amino acids by 7-fluoro-4-nitrobenzo-2-oxa-1,3-diazole. *Anal. Chem. Acta* **130**, 377–383.

Johnson L., Lagerkvist S., Lindroth P., Ahnoff M., and Martinsson K. (1982) Derivatization of secondary amino acids with 7-fluoro-4-benzofurazanyl ethers. *Anal. Chem*, **54**, 939–942.

Jones B. N. and Gilligan J. P. (1983) *o*-Phthaldialdehyde precolumn derivatization and reversed-phase high-performance liquid chromatography of polypeptide hydrolysates and physiological fluids. *J. Chromatogr.* **266**, 471–482.

Jones B. N., Paabo S., and Stein S. (1981) Amino acid analysis and enzymatic sequence determination of peptides by an improved *o*-phthaldialdehyde precolumn labeling procedure. *J. Liq. Chromatogr.* **4**, 565–586.

Joseph M. H. and Davies P. (1983) Electrochemical activity of *o*-phthalaldehydemercaptoethanol derivatives of amino acids. Application to high-performance liquid chromatographic determination of amino acids in plasma and other biological materials. *J. Chromatogr.* **277**, 125–136.

Katz R. I., Chase T. N., and Kopin I. J. (1969) Effect of ions on stimulated-induced release of amino acids from mammalian brain slices. *J. Neurochem.* **16**, 961–967.

Larson B. R. (1982) A model system for mass spectrometric sequencing of *ortho*-phthalaldehyde peptide derivatives. *Life Sci.* **30**, 1003–1012.

Lee S. and Drescher D. G. (1978) Fluorometric amino acid analysis with *o*-phthaldialdehyde (OPA). *Int. J. Biochem.*, **9**, 457–467.

Lindroth P. and Mopper K. (1979) High performance liquid chromatographic determination of subpicomole amounts of amino acids by precolumn fluorescence derivatization with *o*-phthaldialdehyde. *Anal. Chem.* **51**, 1667–1674.

Martin del Rio R., Orensanz Munoz L. M., and DeFeudis F. V. (1977) Contents of β-alanine and γ-aminobutyric acid in regions of rat CNS. *Exp. Brain Res.* **28**, 225–227.

Neadle D. J. and Pollitt R. J. (1965) The formation of 1-dimethylamino-naphthalene-5-sulphonamide during the preparation of 1-dimethylamino-naphthalene-5-sulphonylamino acids. *Biochem. J.* **907**, 607–608.

Norris P. J., Smith C. C. T., de Belleroche J., Bradford H. F., Mantle P. G., Thomas A. J., and Penny R. H. C. (1980) Actions of tremorgenic fungal toxins or neurotransmitter release. *J. Neurochem.* **34**, 33–42.

O'Fallon J. V., Brosemer R. W., and Harding J. W. (1981) The Na$^+$, K$^+$-ATPase: A plausible trigger for voltage-independent release of cytoplasmic neurotransmitter. *J. Neurochem.* **36**, 369–378.

Roth M. (1971) Fluorescence reaction for amino acids. *Anal. Chem.* **43**, 880–882.

Roth M. (1978) Fluorimetric determination of free hydroxyproline and proline in blood plasma. *Clin. Chim. Acta* **83**, 273–277.

Sandberg M. and Corazzi L. (1983) Release of endogenous amino acids from superior colliculus of the rabbit: In vitro studies after retinal ablation. *J. Neurochem.* **40**, 917–921.

Sandberg M. and Jacobson I. (1981) β-alanine, a possible neurotransmitter in the visual system. *J. Neurochem.* **37**, 1353–1356.

Sandberg M. and Lindstrom S. (1983) Transmitter amino acids in the dorsal lateral geniculate nucleus of the cat — collection in vivo. *J. Neurosci. Meth.* **9**, 65–74.

Sandberg M., Jacobson I., and Hamberger A. (1982) Release of endogenous amino acids in vitro from the colliculus and the hippocampus. *Prog. Brain Res.* **55**, 157–166.

Sano I. (1970) Simple peptides in brain. *Int. Rev. Neurobiol.* **12**, 235–263.

Schon F. and Kelly J. S. (1975) Selective uptake of (^{3}H)-β-alanine by glia: Association with the glial uptake system for GABA. *Brain Res.* **86**, 243–257.

Simons S. S. and Johnson D. V. (1976) The structure of the fluorescent adduct formed in the reaction of o-phthaldialdehyde and thiols with amines. *J. Am. Chem. Soc.* **98**, 7098.

Simons S. S. and Johnson D. V. (1977) Ethanethiol: A thiol conveying improved properties to the fluorescent product of o-phthaldialdehyde and thiols with amines. *Anal. Biochem.* **82**, 250–254.

Srinivasan V., Neal M. J., and Mitchell J. F. (1969) The effect of electrical stimulation and high potassium concentrations on the efflux of (^{3}H)-γ-aminobutyric acid from brain slices. *J. Neurochem.* **16**, 1235–1244.

Stein W. H. and Moore S. (1954) The free amino acids of human blood plasma. *J. Biol. Chem.* **211**, 915–926.

Umagat H., Kucera P., and Wen, L.-F. (1982) Total amino acid analysis using pre-column fluorescence derivatization. *J.Chromatogr.* **239**, 463–474.

Ungerstedt U., Herrera-Marschitz M., Jungelins U., Ståhle L., Tossman U., and Zetterström T. (1982) Dopamine synaptic mechanisms reflected in studies combining behavioural recordings and brain dialysis, in *Advances in Dopamine Research* (Kohsake et al., eds.) **37**, pp. 219–231.

Venema K., Leever W., Bakker J.-O., Haayer G., and Korf J. (1983) Automated precolumn derivatization device to determine neurotransmitter and other amino acids by reversed-phase high-performance liquid chromatography. *J. Chromatogr.* **260**, 371–376.

Watkins J. C. and Evans R. H. (1981) Excitatory amino acid transmitters. *Ann. Rev. Pharmacol.* **21**, 165–204.

Chapter 6

Radioreceptor Assays for Amino Acids and Related Compounds

JOHN WILLIAM FERKANY

1. Introduction

The recognition of the significance of amino acids as mediators of chemical neurotransmission in the mammalian central nervous system has grown extensively in recent years, and no fewer than 16 of these simple compounds have been proposed to function as excitatory or inhibitory neurotransmitters in the brain (Enna, 1978a; Watkins and Evans, 1981; Toggenburger et al., 1982; Recasens et al., 1982; Iwata et al., 1982; Baba et al., 1983; Enna and Gallagher, 1983; Foster and Fagg, 1984). As the role of amino acids in neurotransmission has become apparent, the search for analytical methods to investigate the neurochemical and functional aspects of amino acids in brain and peripheral tissues has gained increasing importance.

One technique that has proven useful in this regard is the competitive binding assay. In this elegantly simple procedure, radiolabeled compounds are incubated with portions of tissue samples containing the neurotransmitter receptor of interest, and the binding of the ligand to specific recognition sites is examined. Because useful competitive binding assays are of high affinity (often, subnanomolar concentrations of the ligand are required to identify the relevant receptor), and because only a limited array of compounds or neurotransmitter-like substances interact with a particular recognition site, it is possible to selectively investigate the biochemical and pharmacological characteristics of a receptor that may be present in only minute quantities in biological samples. Aside from being a necessary first step in the isolation of receptors for neurotransmitters or neuromodulators, competitive binding assays have provided substantial information concerning the development, regulation, and distribution of neurotransmitter receptors in a variety of organs, tissues, and species. Moreover, by examining the characteristics of receptors in both normal and pathological conditions, valuable information has been gathered concerning the functional and physical abnormalities that accompany a number of neuropyschiatric disorders.

The use of competitive binding assays to explore the fundamental characteristics of neurotransmitter receptors is exciting, but these techniques have additional important applications. Because the binding of a radioligand to a receptor is highly specific (analogous to the binding of an antibody to an antigen), only a limited number of compounds will inhibit the ligand/receptor interaction. Furthermore, because the reaction between the receptor and compounds that recognize the "active site" of the receptor are usually of high affinity, competitive binding assays can be adapted to detect the small amounts of biologically active substances that are present in tissue samples. When used in this manner, competitive binding assays are generally referred to as "radioreceptor assays" (RRA) and these techniques have provided a simple, sensitive, rapid, and inexpensive method to assay for the presence of physiologically or pharmacologically important substances in tissues. Because RRAs provide a sensitivity comparable to, and in some instances, greater than other analytical procedures to identify and quantitate chemicals and neurotransmitters, the popularity of the technique has grown considerably since first introduced by Lefkowitz and coworkers (1970) some 15 years ago.

Beyond the uses cited above, competitive binding assays have become important in an additional area of neurobiological research—the search for endogenous substances that may function as neurotransmitters or otherwise exhibit neuroactive, receptor-mediated actions. For instance, several compounds of pharmacological importance, including the barbituates, benzodiazepines, and the neurotoxin, kainic acid, have been reported to interact with specific binding sites in the CNS (Leeb-lundberg et al., 1981; Olsen and Leeb-lundberg, 1981; Braestrup and Squires, 1977; Mohler and Okada, 1977; London and Coyle, 1979), and such observations have lead to speculation that the CNS actions of these substances are mediated by receptors that respond to unidentified neurotransmitters or neuromodulatory substances. The same properties that make competitive binding assays useful as RRAs, confer on these methods a unique role in the search for novel, neurotransmitter-like substances. The value that competitive binding assays have achieved in this regard is exemplified by the importance the methods acquired during the initial search for the endogenous opiate-like factors in brain (Terenius and Wahlstrom, 1974, 1975; Pasternak et al., 1975; Teschenaker et al., 1975; Hughes et al., 1975; Simantov and Snyder, 1976a, b, c; Garcin and Coyle, 1976).

The focus of the current chapter is to briefly review the fundamental aspects of the competitive binding assay, particularly as these relate to the use of ligand binding techniques such as RRAs. Emphasis has been placed on practical considerations that should be appreciated prior to attempting to use such methods as analytical techniques. An attempt has been made to discuss both the advantages and disadvantages of RRA as an analytical technique and whenever practical, specific examples of procedures that employ amino acids or amino acid analogs as ligands are discussed. It is recommended that readers desiring additional information concerning competitive binding assays in general, or RRAs in particular, should consult the original articles cited, or any of the several reviews that are available elsewhere on these subjects (Enna, 1978b, 1980a).

2. Competitive Binding Assays

2.1. Saturation

Universally, competitive binding assays are characterized by two important features—saturability and specificity—each of which must be demonstrated to secure the validity of the technique and to document that the ligand/receptor interaction examined in vitro is related to the physiologically important events that occur in vivo. These aspects of ligand binding are discussed respectively in the following two sections of this chapter.

The principle that a receptor can be "saturated" by a ligand dictates that, following the incubation of the tissue in the presence of a certain concentration of radioactive tracer, no further increase should be observed in the amount of tracer bound to the tissue that cannot be displaced by a large excess of the unlabeled ligand or a similar compound that interacts in a competitive manner with the binding site of the receptor. To establish this property of ligand binding, saturation experiments are conducted and, these investigations are among the first to be performed when examining the interaction of a novel radiolabeled compound with specific recognition sites located on tissue suspensions. Beyond demonstrating a criterion important to documenting the biological relevance of the binding phenomenon, saturation analysis yields information about the equilibrium binding constants of the reaction that are crucial to further investigations.

In principle, the construction of "saturation curves" is straightforward, and in these experiments a fixed concentration of the tissue preparation (and thus, a fixed concentration of the receptor of interest) is incubated in the presence of increasing concentrations of the ligand. Since in many instances the dissociation constant (K_d) for the binding reaction is unknown, initial experiments may involve as many as three to four log units of concentration of the tracer. Once the affinity of ligand for the receptor has been established, a more restricted range of ligand concentrations are employed in routine saturation analysis. However, even if the affinity of the ligand for the binding site can be anticipated in advance of the experiment, concentrations of the tracer from at least tenfold less to tenfold greater than the K_d are routinely employed to insure the accuracy and completeness of the binding data.

Variations in the concentration of the ligand can be achieved by either of two approaches, both of which yield identical information. If the affinity of the tracer for the receptor in question is expected to be "high" and the concentration of the ligand required to occupy one-half of the binding sites (i.e., the K_d of the reaction) is in the nanomolar range, it is possible to increase the amount of radiolabel present in the incubations. On the other hand, if the K_d of the interaction is lower (>100 nM), a dilution/displacement method may be used. In this instance, both the concentration of the tissue and of the radiolabel are fixed and increasing amounts of the unlabeled form of the ligand are added to the incubations. The high specific radioactivity (low actual concentration), coupled with the relatively low affinity of some important binding reactions (notably, those involving amino acid-linked receptors), frequently makes the latter procedure the method of choice in saturation experiments.

The portion of ligand binding that is nonspecific and represents the attachment of the tracer to biologically irrelevant sites is routinely determined by conducting some incubations in the presence of a large excess of a nonradioactive inhibitor that interacts competitively at the binding site of the receptor. Since the number of potential nonspecific binding sites is large, sufficient sites exist to accommodate the binding of some tracer, even in the presence of excess inhibitor. Thus, the nonspecific ("blank") binding of a ligand is seldom negligible, although in most useful assays the ratio of total binding to blank binding is at least two to one. In any e vent, the difference between the amount of ligand bound in the presence and absence of the nonradioactive displacer is referred to as "specific binding" and presumably represents the binding of the ligand to biologically important receptors. On theoretical grounds (Cuatrecasas and Hollenberg, 1976; Burt, 1980), it has been argued that the concentration of inhibitor used to define nonspecific binding should not be in excess of 100-fold of the K_d for the ligand/receptor interaction. However, in practice this seems less critical. For instance, in the receptor binding assay for [^{3}H]-gamma-aminobutyric acid ([^{3}H]-GABA), concentrations of unlabeled GABA 1000- to 1,000,000-fold greater than the affinity of [^{3}H]-GABA for its recognition site in brain fail to elicit a detectable change in the amount of ligand nonspecifically bound to tissue preparations.

A typical saturation curve constructed by increasing the amount of radioactive tracer present in the incubations is shown in Fig. 1 for a hypothetical compound, [^{3}H]-ligand. As is evident from the example, the total amount of the ligand bound to the tissue increases in a curvilinear fashion until an inflection is reached at which additional binding of the tracer is linear with tracer concentration. By subtracting the value for nonspecific ligand binding from the total amount of ligand binding, the component of binding attributable to an interaction of the ligand with the receptor is determined. The parallel increase in both total and blank binding beyond the inflection point means that no additional increase in specific binding will occur, and this point of the curve represents the concentration of the ligand necessary to saturate the available receptors.

Since the equilibrium binding constants (K_d; B_{max}) are difficult to determine from curvilinear graphs of the type shown, data are routinely transformed to produce linear plots. The most popular of these transformations utilized in receptor binding studies is that formulated by Scatchard (1949), and application of this method to the data shown in Fig. 1 yields a monophasic curve (Fig. 2) that could be interpreted to suggest that [^{3}H]-ligand binds to only a single population of receptors under the conditions of the assay. In practice, linear transformation of binding data frequently results in two or more lines, indicating the existence in the tissue of multiple recognition sites for the tracer.

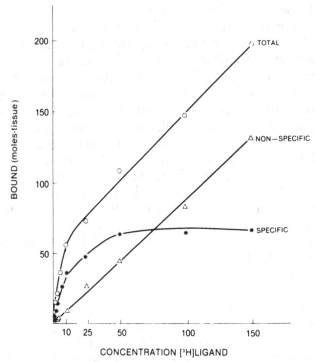

Fig. 1. Hypothetical equilibrium binding analysis for the ideal tracer, (^{3}H)-ligand showing total, nonspecific, and specific ligand binding. The curves were constructed by incubating a fixed amount of tissue in the presence of increasing concentrations of (^{3}H)-ligand with or without a large excess of unlabeled ligand. The inflection in the total binding curve represents the concentration of tracer required to occupy all available receptors.

Scatchard analysis of saturation curves provides two valuable pieces of information concerning the binding reaction. These are respectively, (a) the apparent number (B_{max}) of receptors present in the tissue that specifically binds the ligand and, (b) the apparent affinity (K_d) of the ligand for the receptor. Although both of these constants are of importance, an accurate estimation of the latter is crucial for RRA, since the K_d defines the ligand concentration required for the optimal precision and sensitivity of the method. Furthermore, when biphasic equilibrium binding curves are evident, a judicious selection of ligand concentration, such that the amount of tracer present in the incubations is less than the K_d of the highest affinity reaction, can limit the reaction to the high affinity population of receptors and the system can be treated as though it were monophasic.

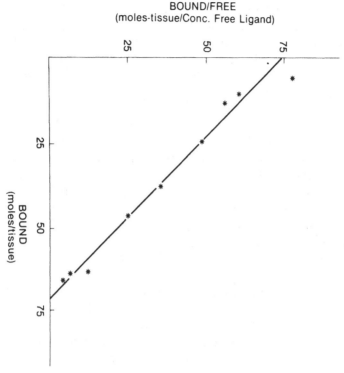

BOUND/FREE
(moles-tissue/Conc. Free Ligand)

Fig. 2. Scatchard transformation of the saturation data depicted in Fig. 1. For this type of analysis only the amount of specifically bound ligand is plotted. The linear equation for these data is $y = -0.1026 + 72$ ($r = -0.98$) and the equilibrium binding constants in arbitrary units are $(X_i) = B_{max} = 72$, $(l/m) = K_d = 9.7$. Monophasic curves of this type would suggest that the tracer bound to a single population of receptors under the conditions of the assay.

2.2. Specificity

Although it is of crucial importance to document that the binding of the tracer to the tissue is saturable, the demonstration of saturability is insufficient to establish the biological relevance of a ligand/receptor interaction. To do this, a second criterion must be fulfilled and it is necessary to document that the reaction of the ligand with the recognition site under examination is "specific." That is, it must be shown that compounds known to possess a biological action at the receptor in vivo should attenuate the binding of the ligand to the receptor in vitro (it should be noted that the meaning of the term "specificity" used here is somewhat different from the definition of the same term as applied in section 6 below). To illustrate, electrophysiological investigations have established that the naturally occurring alkaloid (+)−bicuculline is a potent and specific in vivo antagonist of the actions of the inhibitory neurotransmitter, GABA (Enna and Gallagher, 1983). Similarly, ligand binding studies using [^{3}H]-GABA have revealed that (+)−bicuculline, but not the physiologically inactive sterioisomer (−)-bicuculline, produces a concentration-dependent inhibition of the specific binding of [^{3}H]-GABA to brain homogenates in vitro (Enna and Snyder, 1977b; Enna and Gallagher, 1983). Inhibition of specific [^{3}H]-GABA binding to brain is also induced by other (+)-bicuculline-sensitive GABA agonists, including muscimol, THIP, aminopropane sulfonic acid, and isoguavacine, and these compounds inhibit the specific binding of the [^{3}H]-GABA in vitro with an order of potency similar to that found by electrophysiological means (Enna et al., 1979). In contrast, agents thought to exert their physiological effects at receptors activated by other neurotransmitters or, even close structural analogs of GABA that interact with the neuronal GABA transport site in brain but that are devoid of (+)-bicuculline-sensitive physiological activity in vivo (guavacine, nipecotic acid, and diaminobutyric acid), fail to attenuate the specific binding of [^{3}H]-GABA to homogenates of brain in vitro (Brehm et al., 1979). In a similar instance, the in vivo electrophysiological actions of glycine on neurons are antagonized by the potent convulsant strychnine, an observation that correlates well with the strychnine-mediated inhibiton of the specific binding of [^{3}H]-glycine to spinal cord preparations. Indeed, strychnine is sufficiently selective in this regard that tritiated strychnine has become the ligand of choice to identify glycinergic receptors in the central nervous system (Young and Snyder, 1973; Snyder, 1975).

In principle, the concept of ligand specificity is simple. In practice, demonstrating that a ligand/receptor interaction is specific can pose formidable problems, and this is frequently attributable to the lack of compounds that are selective for a given recognition site in the tissue being examined. In particular, this phenomenon has been frustratingly apparent when the binding of excitatory amino acids and their analogs to brain has been examined. For instance, although it is widely accepted that L-glutamate is an excitatory neurotransmitter in mammalian brain (Foster and Fagg, 1984; Watkins and Evans, 1981), investigations of the site(s) labeled by [^{3}H]-L-glutamate have been hindered by a lack of compounds that selectively antagonize the actions of the amino acid in vivo and inhibit the specific binding of [^{3}H]-L-glutamate to brain membranes in vitro. Furthermore, some agonists thought to act in vivo at receptors fundamentally distinct from those activated by glutamate are potent inhibitors of the specific binding of [^{3}H]-L-glutamate to brain membranes (Watkins and Evans, 1981; Foster and Fagg, 1984; Ferkany and Coyle, in press).

Apparently highly specific interactions of a ligand with a binding site can paradoxically delay the confirmation that specific binding of the ligand occurs at a physiologically relevant site. Thus, numerous biologically active peptides and hormones, including bombesin, cholecystokinin, substance P, neurotensin, and somatostatin, bind in a saturable manner to recognition sites located in multiple tissues. The interpretation of these findings is rendered difficult however, since only the parent compounds were found to be potent inhibitors of ligand binding (Kitabgi et al., 1977; Moody et al., 1978; Nakata et al., 1978; Williams, 1983; Aguilera and Parker, 1982).

2.3. Additional Considerations

Beyond the features of saturability and specificity, competitive binding assays are characterized by several additional properties, all of which must be routinely examined in the course of establishing the validity of a ligand binding technique. Since the reader should be cognizant of at least some of these considerations prior to attempting to employ ligand binding techniques as analytical tools, a brief mention of the more important attributes of the ligand/receptor interaction is made here. More detailed discussions of these considerations can be found elsewhere (Enna, 1980b; Burt, 1980; Yamamura et al., 1978; Cuatrecasas and Hollenberg, 1976).

Established in vitro competitive binding assays for neurotransmitter-like ligands are optimal at physiological pH, an observation that most likely reflects the conditions in which the receptor functions in vivo. Although the acceptable pH for some ligand binding assays may be fairly broad (6.8–7.7), other methods can require that a narrow range of H$^+$ ion concentration be maintained. Routinely, the specific binding of a ligand to receptors located on membranes derived from biological sources declines dramatically when pHs of less than 6.5 or greater than 8.0 are achieved.

In contrast to the general predictability for the pH of the medium in which to conduct the ligand binding assay, the temperature at which the maximum specific binding of a ligand to the tissue will occur must be determined empirically, although most

competitive binding assays are conducted between the temperatures of 4 and 37°C. In any event, specific binding of a ligand that
is apparently saturable and that occurs at extremes of either temperature or pH should be suspect. Indeed, a simple and widely
used test to reinforce the proposal that a tracer binds to a biologically relevant site on membrane preparations is the demonstration that most, if not all, of the displacable binding of the tracer to
the tissue is eliminated following treatments that result in the de-
naturation of proteins.

Like enzymatic reactions, the binding of a ligand to receptors
should proceed over a finite period of time. As is the case with the
determination of the optimal temperature at which to conduct a
binding assay, the period of time required for a particular ligand/
receptor interaction to achieve equilibrium must be determined
by experiment. In most assays involving neurotransmitter receptors, the interaction of the ligand with the receptor reaches equilibrium within 2 h (in contast to radioimmunoassays that may require up to 24 h to obtain a steady state condition). At
equilibrium, the specific component of the ligand/receptor interaction should be reversible. Following the addition to the reaction
of a large excess of a site-directed competitive inhibitor, or following the "infinite" dilution of the radiolabeled tracer by the addition of a large quantity of ligand-free buffer, the amount of tracer
bound to "receptors" should approach or reach values for
nonspecific binding. In fact, the determination of both the for-
ward (association) and backward (dissociation) kinetic constants
for a particular competitive binding reaction provides an
alternative to the methods of linear transformation to estimate the
dissociation constant of the particular reaction.

The specific binding of a ligand to tissue should be linear
with respect to the amount of tissue included in the individual incubations. That is, as the concentration of the tissue in the incubation is altered, the amount of displacable binding of the ligand
should vary proportionately to reflect the increase or decrease in
the concentration of receptors in the assay. Although frequently
considered to be a trivial experiment, documentation of the relationship between the specific binding of the ligand and tissue
concentration is of the utmost importance. Thus, if this relationship is not strictly linear, interassay variations in tissue concentration may result in erroneous estimates of equilibrium binding
constants, or of the potency of various compounds to inhibit ligand binding. Furthermore, when the specific binding of a tracer
is not linear with receptor concentration, the possibility of positive or negative ligand/receptor interactions or the possibility that

the reaction is an artifact of the method, should be immediately entertained. Finally, most ligand binding assays are linear over a certain and limited range of tissue concentrations (Cuatrecasas and Hollenberg, 1976) and, when conducting a competitive binding assay it is imperative to adjust the amount of tissue included in the incubations accordingly in order to maintain acceptable conditions for the assay.

The amount of the ligand specifically bound to the tissue will vary with respect to the tissue that is chosen as a source of receptors. Even within a single tissue like brain, displacable binding may vary considerably among different brain regions. Although such variations can be caused by differences in the affinity (K_d) of the receptor for the ligand, these most often reflect regional differences in the density (B_{max}) of that receptor. For example, the specific binding of the muscarinic cholinergic receptor antagonist [^{3}H]-quinuclidinyl benzolate is high in the corpus striatum and hippocampus and lower in the cerebellum, a pattern that correlates well with the regional distribution of muscarinic cholinergic innervation in the central nervous system (Snyder et al., 1975).

The binding of a ligand can also be influenced by the age and species of the animal from which the receptors are obtained (Slevin and Coyle, 1981), as well as by a variety of ions, enzymes, detergents (Enna and Snyder, 1977b; Fagg et al., 1982; Baudry et al., 1983), or other neurotransmitters and/or neuromodulators (Tallman et al., 1978; Massotti et al., 1981; Ferkany et al., 1984).

Although by no means complete, the above discussion serves to illustrate some of the multiple issues confronted when developing and characterizing a competitive binding assay in its initial stages. For the individual interested in attempting a ligand binding assay for the first time, it is prudent to become familiar with the various factors that influence the particular technique and to rigorously adhere to published methods.

2.4. Artifacts

Numerous artifacts that can give rise to misleading or erroneous data can accompany the in vitro receptor binding assay, and some of the more prominent of these are discussed below. In most instances, careful attention to detail and an immediate suspicion of interassay variations will suffice to avoid the most frequently encountered of these effects.

One of the most common problems associated with competitive binding assays is the decomposition of the radidoligand during shipping or prolonged storage. Although this is less troublesome with radiolabeled amino acids, it can be particularly prominent with labile compounds like the biogenic amines. Even if the parent compound is stable in solution, it should never be assumed that the radiochemical behaves in a similar fashion. Thus, the excitatory amino acid antgonist, 2-amino-7-phospho-noheptanoic acid (APH), can be stored (−20°C) for considerable periods without loss of biological activity, whereas [^{3}H]-APH decomposes by tritium exchange with the solvent under identical conditions (Chapman et al., 1983; Ferkany and Coyle, submitted).

When the decomposition of the ligand goes undetected, several events may occur, all of which adversely affect the characteristics of the binding assay. If the product of decomposition retains the radioactive portion of the molecule and has a detectable affinity for the tissue, spuriously high amounts of nonspecific binding will be apparent. The resulting decrease in the ratio of total to nonspecific binding that occurs in this instance will adversely affect both the sensitivity and precision of the binding assay. On the other hand, if the decomposition product does not bind to the tissue or does not retain the radiolabel, decreases in both the total and nonspecific components of binding may be evident. In either instance, errors in the estimation of the equilibrium or kinetic constants for the ligand/receptor interaction will arise since the true concentration of the tracer in the assay is unknown. Determination of the radiochemical purity of a ligand upon receipt and at periodic intervals is prudent, and most commercial suppliers provide details of chromatographic procedures that can be used for this purpose. When decomposition of the ligand is excessive (>5–10%), repurifying or discarding the tracer is required.

Although less frequently encountered, instances of substantial affinity of a ligand for glass fiber filters and glass or polypropylene centrifuge tubes are not unknown (Innis et al., 1981; Cuatrecasas and Hollenberg, 1976). As is the case for some types of tracer decomposition, a ligand/materials interaction results in an increase in the amount of nonspecific binding that, if sufficient, may obscure the component of binding specific for the receptor. This latter artifact is easily detected by conducting incubations in the absence of a receptor source and, if present, can often be eliminated by using different filters, changing from a filtration to a centrifugation technique to terminate the reaction, or changing the ionic composition of the incubation buffer. In difficult cases, pretreatment of the equipment and/or materials used in the assay with special solvents (e.g., 0.1% v/v polyethylimine or 0.1% bovine serum albumin) may be required (Innis et al., 1981; Luthin and Wolfe, 1984).

Decreases in the total, and thus specific, binding of the ligand may also result from a failure to rigorously maintain the optimal binding conditions with respect to temperature, pH, and equilibration time. Since these conditions are generally established in preliminary studies, they only rarely contribute to interassay variations. A more common source of variations, particularly when competitive binding assays are adapted for use as RRAs, is the introduction to the assay of contaminants that are present in the sample being examined. Thus, the brain receptors labeled in vitro by [^{3}H]-GABA can be studied using either freshly prepared or previously frozen tissue specimens if it is realized that in the former instance, the ligand has a substantial affinity for the GABA transport site when Na$^+$ ions are present (Enna and Snyder, 1977b). The pharmacologic specificity of the transport site is radically different from that of the postsynaptic GABA receptor. If the object of a particular study were to detect or quantify compounds that act at the neuronal GABA receptor, it would be inappropriate to employ sample preparation techniques that might introduce Na$^+$ to the incubation. Similarly, the specific binding of [^{3}H]-L-glutamate to brain membranes is altered by millimolar concentrations of sodium, potassium, calcium, and chloride (Baudry et al., 1981a, b; 1983; Fagg et al., 1982; Foster and Fagg, 1984), and caution must be exercised with this assay to ensure the exact composition of the incubation medium.

Conversely, it should be noted that specific and competitive binding of some ligands can be detected only in the presence of one or more ionic species. If buffers containing mixes of ions are required, these should be prepared daily to avoid the formation of precipitants or ionic complexes. Some frequently employed buffering agents, notably citric acid, are known to weakly chelate divalent cations, and such combinations should be avoided if at all possible.

Most competitive binding assays for neurotransmitter receptors are terminated by centrifugation to produce a pellet containing the bound radioactivity, or by trapping the tissue on a suitable support using vacuum filtration. In either instance, residual and unbound ligand is removed by rapidly and superficially washing the pellet or filter with small amounts (5–10 mL) of ice-cold buffer. A common and avoidable source of assay variation that plagues those unfamiliar with binding methods evolves from inattention to the length of time the tissue is rinsed following termination of the reaction. Because washing the tissue with ligand-free buffer represents an infinite dilution of the tracer, some dissociation of specifically bound radioactivity will always take place even when the rinse is extremely rapid. When rinsing is excessive or uneven, a sizable variation in the amount of specific binding can result. In this regard, Bennett (1979) has calculated that the allowable separation time (<10% dissociation of the specifically bound component of the interaction) is 10 s if the K_d of the interaction is 10 nM). However, if the affinity of the ligand for the receptor is tenfold lower (100 nM), the 10% dissociation period shrinks to <0.1 s. Clearly, when assays use ligands with affinities for the receptor of less than 10^{-8} M, careful attention to the washing procedure is warranted.

Caution should also be exercised in selecting a tissue to be used as a source of receptors for a particular binding assay, and although many procedures begin with whole brain, this can be an unacceptable practice. The specific binding of [^{3}H]-spiperone to striatal dopamine receptors, for instance, forms the basis of an RRA to detect and quantify the neuroleptic content of human serum (Creese and Snyder, 1977). While [^{3}H]-spiperone specifically labels dopamine receptors in the striatum (Fields et al., 1977), in other areas of brain, serotonergic receptors may be labeled as well (Leysen et al., 1978; List and Seeman, 1981; Peroutka and Snyder, 1983), and the use of whole brain as a source of receptors for the neuroleptic RRA would be ill-advised. Ligands such as [^{3}H]-kainic acid have been shown to interact with a single population of receptors in some areas of brain, but identify two or more pharmacologically similar, but kinetically distinct, sites in others (London and Coyle, 1977). When multiple and pharmacologically distinct receptors for a single ligand are present throughout the brain, as is apparent for the β-adrenergic systems, the ratio of distinct recognition sites may vary among regions (Pitman et al., 1980). In either of these latter cases, both the specificity and sensitivity of an RRA would be altered by the indiscriminate use of different tissue preparations. If conditions dictate a change in the tissue to be used as a receptor source for an RRA, it is wise to reassess the basic characteristics of the system before proceeding.

Finally, it is not unknown for those attempting an RRA for the first time to succumb to the urge to increase the protein and/or ligand concentrations to excessive amounts on the assumption that "more must be better." Such temptations should be resisted at all costs, and although these manipulations may substantially increase the amount of ligand bound in the assay, they can produce equally detrimental effects. Large increases in protein concentration may place the assay beyond acceptable limits with respect to tissue linearity, or may decrease the ratio of total to nonspecific binding, effecting a loss in the precision of the technique. Increasing the ligand concentration in excess of the K_d of the reaction decreases the sensitivity of the method as an RRA (*see* section 5), and may result in the appearance of pharmacologically distinct and/or biologically unimportant binding.

3. Applications

3.1. Characterization of Receptors

The first and foremost use of competitive binding assays is to define the biochemical and pharmacologic characteristics of receptors for neurotransmitters or neuromodulators in a particular tissue. By identifying these and other properties of receptors in vitro, it is possible to gain an insight into the physiological role played by different substances in a particular organ. Furthermore, the comparison of neurotransmitter or hormonal receptors in the normal and diseased states has substantially increased the understanding of the functional deficits that occur in many pathological conditions. When combined with a broader understanding of a particular disorder, this information can lead to the development or recognition of suitable therapeutic strategies for a given disease.

Although this application of competitive binding assays is exciting, even a limited discussion of specific cases is beyond the scope of this chapter. It is recomended that interested readers should consult any of the excellent reviews that are available on selected topics in the literature.

3.2. Radioreceptor Assays

Also widespread is the application of competitive binding techniques as RRAs and, in this instance, the goal is to detect and quantify trace amounts of biologically active substances in tissues and fluids. The utility of competitive binding assays in this regard arises from two observations, and like radioimmunoassays, RRAs are (a) sensitive and detect minute amounts of the compound of interest and, (b) are selective, detecting only the compound of interest.

The principles of RRAs are straightforward and similar to those for radioimmunoassays. In the RRA, the quantity of a fixed amount of radioligand that is specifically bound to a receptor is a function of the concentration of inhibitor present in the incubation medium. When standard curves are constructed that relate the amount of tracer specifically bound to the tissue in the presence of known amounts of a competitive inhibitor, it becomes a matter of comparison to quantitate the substance of interest. Using this method, RRAs for several biologically active substances have been developed and some of the more widely used procedures are methods that detect and quantify neuroleptics (Creese and Snyder, 1977), tricyclic antidepressants (Innis et al., 1979), opiates (Childers et al., 1977; Simantov et al., 1977), β-adrenergic antagonists (Innis et al., 1978), the inhibitory amino acid neurotransmitter GABA (Enna and Snyder, 1976; 1977a; Ferkany et al., 1978), and calcium channel antagonists (Gould et al., 1983).

3.3. Detection of Novel Receptor Active Substances

The final use of competitive binding assays is the detection of novel neurotransmitter-like or neuromodulatory substances. In this application, tissue extracts are isolated using standard separation techniques and these are screened for the presence of factors that displace a ligand specifically bound to a particular receptor. Isolated fractions of tissue extracts that inhibit the binding of a ligand to its recognition site can then be further purified in attempts to identify the active substance in a homogenous form. By applying competitive binding assays in this manner, several laboratories found these techniques useful during the initial search for the endogenous opiate-like substance(s) in the brain, and receptor binding assays played an important part in the isolation and subsequent identification of the endorphins and enkephalins in the CNS (Terenius and Wahlstrom, 1974; 1975; Pasternak et al., 1975; Teschenaker et al., 1975; Hughes et al., 1975; Simantov and Snyder, 1976a,b,c; Garcin and Coyle, 1976).

Although the practical aspects of the separation and isolation chemistry involved in the detection of novel compounds of neurochemical interest remain complex, the availability of competitive binding assays for a varietiy of receptor types has made the detection of neuroactive compounds remarkably simple. Like RRAs, these procedures are of high specificity and sensitivity and afford a means to rapidly examine a large number of extracted fractions in a short period of time. When compared to the more tedious physiological assays that were formerly the only means of detecting novel biologically active substances, these properties lend the competitive binding assay a particular advantage and attractiveness.

Because the principles of competitive binding assays for this last application are identical to those of the RRA, they will be discussed together in the following sections. For simplicity, both applications of ligand binding techniques are referred to as "RRAs."

4. Assay Procedures

4.1. Equipment and Reagents

The equipment and reagents required to successfully perform RRAs are minimal and generally present in any laboratory routinely engaged in biochemical research.

An essential requirement is the refrigerated centrifuge capable of generating forces from approximately 1000–50,000g. Appropriate and dependable instruments for this purpose include the DuPont Sorvall RC-5 and Beckman J2-21 series of instruments. Because the centrifuge is used both in the preparation of tissues and, not infrequently, to terminate the binding reaction, it should be equipped with rotors capable of handling a small number of large samples (≤50 mL) and multiple samples of smaller volumes.

Although some receptor binding assays, including those for the amino acids (glutamate, aspartate, GABA), are terminated by centrifugation, assays for the biogenic amines, opiates, benzodiazepines, and muscarinic cholinergic receptors, to name a few, are halted by rapid vacuum filtration. For this purpose, both commercially supplied and custom-built manifolds are available that consist of multiple chambers to support suitable filters and a solvent tank to collect waste. Recently, Brandell (Gaithersburg, MD) has introduced a modified cell harvester for use in competitive binding assay. This semiautomated unit provides some advantages over manual devices because the reaction for every sample is terminated simultaneously and the rinsing of the filters is identical in all chambers and between assays.

In addition to the refrigerated centrifuge and the filtering manifold, other essential equipment required for RRAs include the liquid scintillation counter to quantify radioactivity, a freezer (≤−20°C) for sample storage, and preparative equipment including a tissue homogenizer or sonic cell disruptor. Beyond these major requirements, all other chemicals, reagents, fluors, and ligands can be obtained commercially from the standard sources.

4.2. Tissue Preparation

An immediate problem that must be overcome when conducting an RRA is the relatively low density of receptors specific for the ligand vis-à-vis other tissue constituents that are available for nonspecific interactions. Indeed, in many competitive binding assays, the amount of ligand specifically bound when all receptors are occupied (saturated) is on the order of fmoles/mg protein. Because in most applications, and particularly for RRAs, nonsaturating concentrations of ligand are employed, the amount of specific binding is proportionally reduced. Thus, for many assays it is routine to attempt some purification and concentration of the relevant binding site in order to increase the proportion of total binding that is specific for the receptors. Although these manipulations rarely produce a homogeneous or even highly purified population of binding sites, even this moderate concentration of receptors can substantially increase the precision of an RRA (Ferkany and Enna, 1980). Because many of the currently available RRAs use brain as a source of receptors, the following discussion will focus on the preparation of this tissue. The principles outlined below, however, are equally applicable when other organs serve as the initial source of binding material.

One of the most frequently used tissue preparations for RRAs is the method developed by Enna and Snyder (1976). Starting with whole brain (or discrete regions of brain, if desired), the tissue is homogenized in 20 vol (w/v) of ice-cold, 0.32M sucrose using a motor-driven Teflon pestle fitted to a glass tube. The homogenate is centrifuged at low speed (1000g; 10 min) to remove large debris, and the resultant supernatant is decanted and recentrifuged (20,000g; 20 min) to produce the crude mitochondrial pellet (P_2) that is enriched in nerve terminals and postsynaptic elements. This pellet is lysed by hypoosmotic shock and mechanical disruption, and the suspension is centrifuged again (8000g; 20 min) to yield a pellet with a dense, tightly adhering core surrounded by a loosely packed "buffy coat." Gentle shaking removes the latter and this is decanted with the supernatant to a clean tube. The homogenate is recentrifuged (50,000g; 10 min) a final time to produce a pellet that is the initial starting material for assay. It should be noted that all of the above steps are routinely performed at 2°C.

Further handling of the tissue varies considerably among assays. To examine the specific binding of [^{3}H]-L-glutamate (Slevin and Coyle, 1981) or [^{3}H]-kainic acid (London and Coyle, 1979), the tissue is simply washed an additional 2–4 times by centrifugation with intermediate resuspensions in fresh buffer, and used directly in the assay. In other assays however, preincubation of the tissue in the absence or presence of detergents or other agents may be required and, for instance, for the [^{3}H]-GABA RRA, treatment of the membranes with Triton X-100 is routine (Enna and Snyder, 1976; 1977a; Ferkany et al., 1978; Toffano et al., 1978; Guidotti et al., 1982).

For many receptor binding assays, tissue may be stored frozen without an appreciable loss of receptor number or receptor affinity and, because of this, batch preparation of tissue is facili- tated. In at least one case, when the binding of the excitatory amino acid antagonist [^{3}H]-2-amino-4-phosphonobutyric acid is studied, freezing the tissue results in a precipitous decline in specific binding of the ligand, and for this assay it is necessary to prepare fresh membranes on a daily basis (Butcher et al., 1983; Monaghan et al., 1983).

In addition to increasing the precision of RRAs, rudimentary purification of the tissue enhances the sensitivity of these meth- ods to detect the compound(s) of interest. This is particularly true if the focus of the assay is amino acids. Many amino acids, including glutamate, aspartate, and GABA are present in brain and other tissues in high concentrations ($\geq 10^{-3} M$). Because the affinity of amino acid ligands for their receptors varies from 10–1000 nM, the presence of large quantities of endogenous dis- placers in the incubation would reduce the specific binding of the ligand by decreasing the apparent affinity of the ligand for the re- ceptor. Even if crude homogenates of brain membranes are used as a source of receptors for an RRA, washing the tissue by serial centrifugation to dilute or remove endogenous inhibitors is required.

In summary, preparation of tissues serves to enhance both the precision and the sensitivity of the RRA technique. Although general similarities in preparative procedures exist among assays, each RRA has been tailored to a specific purpose and the tech- niques established during the course of developing the assay should be adhered to with rigor.

4.3. Sample Preparation

When the goal of an RRA is to detect or quantify a known sub- stance in biological specimens, the preparation of the sample con- taining the substance can be quite minimal, especially when com- pared to the preparative steps that are sometimes required prior to using more sophisticated analytical techniques. For example, the specific binding of [^{3}H]-GABA to brain membranes forms the basis of an RRA to quantitatively determine the amount of GABA in blood, brain, and cerebrospinal fluid (Enna and Snyder, 1976; 1977a; Ferkany et al., 1978) and, in the latter instance, aliquots of cerebrospinal fluid may simply be added directly to the binding assay. To determine the amount of GABA in either blood or brain, known amounts of tissue are homogenized in a weak acid and the precipitate is removed by centrifugation and the acid extract neu- tralized. Aliquots of this aqueous preparation containing the extracted amino acid are then added to the assay in lieu of an equal amount of the incubation buffer. To precipitate the proteins in the tissues, perchloric acid has proven useful, since titration with KOH yields an insoluble salt that is easily removed by filtration or centrifugation of the aqueous extract.

As with other analytical procedures, the complexity of the sample preparation that is required will vary according to both the binding assay that forms the basis of the RRA and the objective of the individual study. To determine the concentrations of dihydropyridine calcium channel antagonists in plasma, aliquots of plasma can be added directly to a [^{3}H]-nitrendipine binding assay (Ferkany and Badshah, unpublished). This simple approach is not applicable, however, to quantitatively assess the plasma concentration of β-adrenergic receptor blockers in serum, because serum contains elements that interfere with the specific binding of [^{3}H]-dihydroalprenolol to adrenergic receptors. In this latter case, it is necessary to extract the compounds of interest from the plasma into an organic solvent and to back extract the drugs from the solvent into a suitable aqueous phase that can be added to the [^{3}H]-dihydroalprenolol binding assay (Innis et al., 1978). The total plasma content of tricyclic antidepressants can be measured by adding plasma directly to an RRA based on the specific binding of the muscarinic cholinergic receptor ligand, [^{3}H]-quinuclidinyl benzolate ([^{3}H]-QNB). However, to differentially measure tertiary amines like imipramine and their secondary aminergic metabolites (desipramine), both of which are present in the plasma of patients receiving imipramine and both of which are equipotent in the [^{3}H]-QNB binding assay, it was necessary to develop a method to separate these drugs from the plasma. To do this, compounds are extracted into a suitable organic phase and the secondary amines chemically reacted to form products that are retained in the organic phase during the back extraction of imipramine to acid. The plasma content of desipramine is then determined by subtracting the measured amount of imipramine in the plasma from the total serum content of antidepressant drugs (Ferkany et al., unpublished).

When RRAs are employed to detect endogenous neuroactive substances of unknown identity, sample preparation becomes highly individualized and must be adjusted according to the binding assay involved, and to the suspected chemical nature of the unidentified compound. Frequently, the handling of tissue extracts is complex and requires a knowledge of separation chemistry and chromatographic techniques (Zaczek et al., 1983; Toffano et al., 1978; Guidotti et al., 1983). However, once extracted or isolated in a useable form, the principles of the RRA that have been discussed above, to detect the presence or to quantitate a particular substance, remain unchanged.

Regardless of the application of the RRA, several considerations that may adversely affect the quality of RRAs are common to sample preparation and deserve brief mention. Because the preparation of the sample containing the compound to be measured often requires the use of acidic, basic, or organic solvents, care should be exercised to ensure the complete removal or neutralization of these components to prevent contamination of the assay. Where aqueous samples have been prepared, serial dilution of the sample will often suffice if the substance of interest is present in sufficient quantities to be detected by the RRA following dilutions of several orders of magnitude. When serial dilution is not possible, the effects of drug-free extract on the characteristics of the RRA must be determined and a correction applied by including an equivalent amount of the extraction medium in reference assays. The contamination of the RRA by seemingly innocuous agents, including ions and proteins like albumin, should not be overlooked, as these substances are known to produce undesirable effects in several binding assays. As with solvent interference, the influence of such agents can be minimized by adequate dilution of the sample or by applying correction factors to the reference assay. In severe instances, when small contaminants continue to degrade the RRA, more stringent steps may be required, including additional chromatographic purification of the extract (Enna and Synder, 1977a).

4.4. Data Analysis

The goal of the RRA is to detect and quantify a substance in a tissue. In this respect, RRAs differ from conventional competitive binding assays that seek to estimate the equilibrium binding constants for the ligand/receptor interaction, and the analysis of the data for RRAs must be adjusted accordingly. Because numerous excellent reviews are available detailing the practical and theoretical considerations for estimating equilibrium binding constants (Cuatrecasas and Holllenberg, 1976; Burt, 1980), only the methods for the interpretation of RRAs will be discussed below.

The cornerstone of the RRA is the standard curve that represents the displacement of specifically bound ligand by known concentrations of an inhibitor. By comparing the reduction in the specific binding of a ligand produced by a sample containing an

unknown amount of inhibitor to the standard curve, a direct estimate of the amount of receptor active substance present in the
sample is obtained.

A typical standard curve for the inhibition by GABA of
[3H]-GABA specifically bound to detergent-treated brain membranes is shown in Table 1. In this example, the total amount of
[3H]-GABA bound in the absence of any inhibitor is 6000 cpm,
whereas nonspecific binding of the ligand, defined as the residual
radioactivity bound to the tissue in the presence of $10^{-3}M$
unlabeled GABA, is 800 cpm. The difference between these
values (5200 cpm) represents the amount of [3H]-GABA that is

TABLE 1

Typical Standard Curve for (^{3}H)-GABA Radioreceptor Assay: Inhibition of
Specifically Bound (^{3}H)-GABA by Known Concentrations of Unlabeled GABA
and by Plasma Extracts[a]

GABA content, nM	Total CPM bound	Specific CPM bound	Percent inhibition	Percent remaining	Unlabeled GABA, pmol/ incubation
[3H] + 5	6000	5200	0	100	—
[3H] + 10^6	800	0	100	0	—
[3H] + 500	1008	208	96	4	1000
[3H] + 250	1320	520	90	10	500
[3H] + 100	1840	1040	80	20	200
[3H] + 50	2566	1766	66	34	100
[3H] + 25	3000	2200	50	50	50
[3H] + 10	4180	3380	35	64	20
[3H] + 5	4960	4160	20	80	10
[3H] + 2.5	5584	4784	8	92	5
[3H] + 1	5792	4992	4	96	2
Sample A 200 μL	2880	2080	60	40	70
Sample B 200 μL	3504	2704	48	52	48
Sample C 200 μL	4700	3900	35	65	16

[a]Rat brain membranes were prepared as described in section 4.2 and incubated in 2 mL
Tris citrate buffer (0.05M, pH 7.1, 2°C, 5 min) in the presence of (^{3}H)-GABA alone (5 nM,
10^6 dpm, SA = 45 Ci/mm) or (^{3}H)-GABA and the indicated concentrations of unlabeled
GABA. The reaction was terminated by centrifugation (50,000g × 10 min, 2°C) and the
pellets were rapidly and superficially rinsed with ice-cold buffer. The tissue was
solubilized in Protosol© (NEN) and 10 mL of Econofluor© (NEN) were added to the tubes.
Radioactivity was determined by liquid scintillation spectrometry.

To determine plasma GABA content, 200 μL of incubation buffer were replaced by 200
μL of a neutral aqueous extract of plasma. Since each mL of plasma was extracted into a
final volume of 2 mL, the absolute amounts of GABA in the initial samples were, respectively: (a) 700, (b) 480, and (c) 160 pmol/mL.

specifically bound to the receptor. To construct the standard curve, nine concentrations of unlabeled GABA ranging from approximately tenfold below to tenfold above the K_d of the ligand/ receptor interaction have been employed. At each of these concentrations of inhibitor, the amount of [^{3}H]-GABA bound to the tissue is determined and the nonspecific binding is subtracted from this value. The resultant number is divided by the total specific binding (5200 cpm) to yield the percent of specific binding of the ligand remaining in the presence of a given concentration of unlabeled GABA. The standard left to right displacement curve for these data (Fig. 3) is drawn on semilogarithmic paper by plotting the percent of specific binding of the ligand remaining at each concentration of inhibitor against the concentration of unlabeled GABA. As with other binding data, inhibition curves of this type are sigmoidal and transformation to a linear format facilitates the interpretation of the data. In practice, this is achieved by graphing the results on log-probability paper and plotting the percent of specific binding remaining against each concentration of inhibitor that was present in the incubation (Fig. 4).

Also shown in Table 1 is the degree of inhibition of specifically bound [^{3}H]-GABA effected by three samples of an

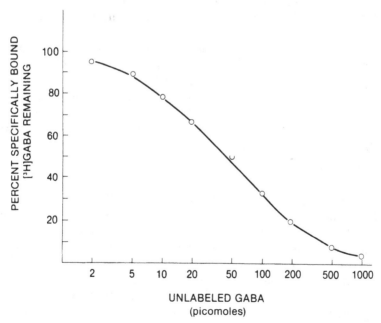

Fig. 3. Displacement of specifically bound (^{3}H)-GABA by unlabeled GABA. Conditions of the assay and the data shown are identical to those given in Table 1.

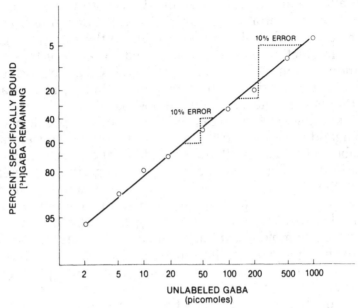

Fig. 4. Log-probability transformation of the inhibition of specifically bound (^{3}H)-GABA by unlabeled GABA. Conditions of the assay and the data shown are identical to those given in Table 1. The error bars represent the possible range of values for a (±) 10% error in the data assuming 80 or 50% inhibition of specifically bound ligand.

aqueous extract of plasma containing an unknown amount of GABA. In this example, the percent of inhibition of specific binding by each sample is 60, 48, and 35%, respectively. Converting these values to the percentage of specific binding remaining (100% − %inhibition), and extrapolating from the standard curve to the ordinate of Fig. 3 reveals that sample B contained 48 pmol of GABA/0.2 mL of extract. Knowing that each 1 mL of plasma yielded 2 mL of extract, then 480 pmol of GABA were initially present in plasma sample B.

Radioreceptor assays produce reliable values when the inhibition of the specifically bound ligand effected by the sample is less than 80% or greater than 20%. Inspection of Fig. 3 also reveals that the most accurate determinations will be obtained when 50% of the tracer specifically bound to the receptor is displaced. Thus, at the midpoint of the logit plot, a 10% error in the precision of the assay yields an estimate of 50 ± 15 pmol GABA, whereas the same error when 80% of the specific binding is inhibited yields a value for GABA of 240 pmol, with a possible range of 110–450 pmol. This property of RRAs dictates that whenever possible, several volumes of a sample should be analyzed in a single assay to ensure that the percentage of inhibition remains on the linear and sensitive portions of the standard curve. When all values are within this range, a determination of the average inhibition of specific binding will provide a value for the concentration of an unknown more accurate than that made on the basis of a single volume of sample.

When the identify of the inhibitory activity present in a sample is unknown (as is frequently the case when RRAs are used to screen for novel substances), the absolute amount of the substance cannot be determined from an RRA. Nevertheless, RRAs are useful to detect the presence of unidentified substances in tissue extracts, to monitor the success of purification techniques used to concentrate or isolate an unknown agent, or to detect the dilution or loss of the inhibitory activity during separation or purification attempts. Even in instances when the identity of the inhibitory activity present in a sample is known, as is the case for the RRA for endogenous opiates (Childers et al., 1977; Simantov et al., 1977), the concentration of the substance may remain indeterminant, since this assay does not discriminate between the various opioid peptides that occur in brain. In either of these latter situations, the difficulty in determining the absolute amounts of the inhibitory activity present in the initial sample is circumvented by expressing the displacement of specifically bound ligand in arbitary units of equivalence relative to a reference compound.

5. Sensitivity

The sensitivity of an RRA is defined as the least amount of substance that the assay can reliably and reproducibly detect. The limit of sensitivity for a particular assay is influenced by at least three factors, including: (a) the affinity of the ligand and the compound of interest for the receptor, (b) the specific radioactivity of the ligand, and (c) the final volume of the incubation. Somewhat surprisingly, the sensitivity of an RRA is inversely related to the concentration of the radioactive tracer included in the assay.

Although it would initially appear that high-affinity ligands would provide more sensitive RRAs, this is not necessarily true, and the sensitivity of an assay is a function of both the affinity of displacer for the receptor and the ratio of the affinities of the displacer and the ligand used to label the receptor. For example, [^{3}H]-muscimol, a potent GABA receptor agonist, has been proposed as a ligand for use in an RRA to detect GABA in biological tissue samples (Bernasconi et al., 1980). Although muscimol has a tenfold higher affinity for the GABA receptor than GABA itself, the use of the former compound as the radioactive tracer produces no increase in the sensitivity of the RRA, because the affinity of the compound being measured (GABA) for the receptor does not change and equivalent concentrations of the amino acid must be present regardless of whether [^{3}H]-muscimol or [^{3}H]-GABA is used as a tracer. Indeed, the lower specific radioactivities of commercially available [^{3}H]-muscimol, relative to [^{3}H]-GABA, may lead to a loss in precision of RRAs using the former compound because the amount of detectable radioactivity bound to the tissue is reduced. In some cases, the sensitivity of an RRA can actually be decreased by employing radioligands with high affinities for a receptor. Utilizing [^{3}H]-spiperone, an RRA has been developed to detect and quantify neuroleptics present in the serum of patients receiving these drugs (Creese and Snyder, 1977). Although the assay relies on the specific binding of [^{3}H]-spiperone to striatal dopamine receptors, it would be ill-advised to attempt to measure endogenous dopamine using this technique because at least $10^{-6}M$ dopamine is required to displace 50% of specifically bound [^{3}H]-spiperone (Fields et al., 1977). A more appropriate strategy would employ [^{3}H]-dopamine as the ligand because under these conditions the IC_{50} (the concentration of drug required to displace 50% of the specifically bound ligand) for dopamine is approximately 8 nM (Burt et al., 1975). As a rule, high-affinity ligands enhance the sensitivity of an RRA only if the affinity for the receptor of the compound being measured is greater than or equal to that of the ligand.

The specific radioactivity of the ligand and the concentration of the ligand added to an RRA are intimately related in determining the sensitivity of a particular assay. Given a twofold increase in the specific radioactivity of the tracer, it is possible within limits to reduce the concentration of the tracer in the assay by one-half with little or no loss in detectable radioactivity. Because the sensitivity of an RRA to detect receptor-active substances is a function of the final concentration of the ligand in the incubation, the use of tracers of the highest specific radioactivity is warranted.

The link between the concentration of the ligand in the assay and the sensitivity of an RRA arises from the relationship shown in Eq. (1):

$$K_I = IC_{50}/[1 + (L/K_d)] \tag{1}$$

where K_I is the apparent inhibition constant for a substance at a particular receptor, IC_{50} is the concentration of the substance required in an assay to inhibit 50% of the specifically bound ligand, L is the concentration of the tracer included in the incubation, and K_d is the affinity of the tracer for the receptor. Rearranging Eq. (1) to emphasize the IC_{50} yields Eq. (2):

$$IC_{50} = K_I[1 + (L/K_d)] \qquad (2)$$

Since both the K_I and the K_d are constants, it is readily apparent the L determines the amount of the substance being measured that must be present in the incubation to inhibit the specific binding of the ligand by 50%. In designing an RRA to detect and quantify trace amounts of biologically active substances in a tissue sample, it is useful to employ the lowest possible tracer concentration in the assay that yields reproducible and accurate analyses.

Using ligands of high specific radioactivity can further increase the sensitivity of an RRA by permitting a reduction in the final volume of the individual incubations. Given a fivefold increase in the specific radioactivity of a tracer, a proportionate reduction in assay volume could theoretically be achieved with little loss in the amount of detectable radioactivity bound to the tissue. The sensitivity of the RRA is increased fivefold because the amount of the substance being measured that must be present in the assay to maintain equimolar concentrations, is reduced fivefold. The absolute reduction in the volume of the assay that can be achieved in practice is limited by several considerations. Most importantly, as the final incubation volume is reduced, the amount of tissue added to the assay decreases. Because this implies that the receptor concentration must also decrease, the quantity of tracer that is specifically bound to the tissue eventually approaches the limits of accurate detection using available liquid scintillation counting methods. Additionally, in some RRAs, contaminants introduced to the assay by the sample exhibit more adverse affects on the characteristics of the ligand/receptor interaction as assay volume is reduced (Enna and Snyder, 1977a).

6. Specificity

The specificity of an RRA defines the ability of the assay to detect only the compound(s) of interest. Since RRAs represent an indirect approach to detect biologically active substances, any compounds or agents that alter the characteristics of the ligand/receptor interaction have the potential to adversely affect specificity. Some of the various instances in which assay specificity can be altered have been discussed in previous sections and these will not be reiterated here. The potential for artifacts that can produce false positive or negative results, however, should always be borne in mind.

The concept of specificity can best be illustrated using two examples. In modifying the [^{3}H]-GABA binding assay to detect and measure the GABA content of blood, it was necessary to exclude the possibility that components of blood other than GABA itself could inhibit the interaction of the ligand ([^{3}H]-GABA) with its receptor. One likely candidate in this regard was imidazoleacetic acid, a naturally occurring metabolite of histamine that is also a potent GABA receptor agonist. To investigate this possibility, a highly specific GC/MS method of quantifying the amount of GABA in blood and plasma was simultaneously developed, and in these investigations, it was demonstrated that the amount of GABA present in blood or plasma was similar when measured by either of the GC/MS or HPLC techniques (Ferkany et al., 1978). Although these results did not exclude the possibility that additional compounds may be present in blood that could interact with the receptor labeled by [^{3}H]-GABA, the data suggested that any such compounds were either eliminated during the extraction procedure developed for this assay or were present in only trace amounts in blood and were of little practical concern with regard to the RRA. In a similar instance, the specific binding of [^{3}H]-L-glutamate to brain membranes was successfully used to identify a novel brain peptide, N-acetylaspartyl-L-glutamate, which appears to interact with a subpopulation of excitatory amino acid-like neurotransmitter receptors in brain. In taking advantage of the [^{3}H]-L-glutamate binding assay to identify this compound, it was necessary to carefully prepare brain homogenates using chromatographic techniques to remove other naturally occurring substances, including L-aspartate, cysteine sulfinic acid, L-aminoadipic acid, L-glutamic acid, and homocysteic acid, all of which are present in brain in substantial concentrations and inhibit the specific binding of [^{3}H]-L-glutamate to brain membranes in vitro (Zaczek et al., 1983).

In some instances, the lack of specificity of a competitive binding assay can be used to advantage when these assays are adapted for RRAs. Although several highly sensitive and accurate GC/MS and HPLC methods are available to measure the amount of various neuroleptics present in the serum of psychiatric patients receiving these drugs, such methods are frequently selective for only a single drug and routinely identify only the parent compound. Because the correlation between plasma levels of the parent compound and patient response is poor (Coyle, 1982), these techniques have been of limited value in designing individual therapeutic strategies. In contrast, a recently developed RRA that relies on the interaction of neuroleptics with the binding site on brain membranes that is labeled by [^{3}H]-spiperone detects not only the original drug administered, but also several of the major metabolites of the neuroleptics. Since these metabolites interact with the dopamine receptor(s), they presumably contribute to the therapeutic response elicited by neuroleptic administration. In fact, the correlation of clinical improvement and serum neuroleptic levels determined by RRA is good (Calil et al., 1979; Tune et al., 1980).

7. Directions

Radioreceptor assays have become popular tools for both clinical and basic research. By taking advantage of these simple methods, a variety of compounds, including the neuroleptics, tricyclic antidepressants, β-adrenergic receptor blocking agents, the endogenous and synthetic opioids, and calcium channel antagonists can be conveniently detected and quantified in biological samples. Because the source of the receptor material for RRAs is biological in nature, any substance that binds in a specific and saturable manner to tissue could theoretically form the basis for an RRA (Table 2).

The major disadvantage of RRAs is the lack of specificity of many of the currently available methods, and this drawback is particularly apparent with competitive binding assays that employ amino acids as ligands. Although well-established procedures exist to examine the fundamental characteristics of L-glutamate and L-aspartate receptors in brain, adaptation of these assays to RRAs to measure glutamate or aspartate would be difficult because endogenous compounds other than these amino acids would almost surely interfere with the specificity of any RRA based upon the competitive binding of [^{3}H]-L-glutamate or [^{3}H]-L-aspartic acid (Slevin and Coyle 1981; Fagg et al., 1983; Foster and Fagg, 1984). For this reason, the only RRA in widespread use for an amino acid detects and quantifies the inhibitory neurotransmitter GABA (Enna and Snyder, 1976; 1977a; Ferkany et al., 1978; Bernasconi et al., 1980).

As more selective ligands for amino acid receptors become available, the adaptation of competitive binding assays to RRAs to detect and measure amino acids should be possible. The excita-

TABLE 2
Competitive Binding Assays Currently Used or of Potential Use as a
Radioreceptor Assay[a]

Ligand	Substance detected	Reference
[³H]-GABA	GABA	Enna and Snyder, 1976
[³H]-Strychnine	Gly	Young and Snyder, 1973
[³H]-Glu	NAAG	Zaczek et al., 1983
[³H]-2-amino-7-phosphono-heptanoic acid	Glu	Ferkany et al., 1984
[³H]-Naloxone	Opiates	Simantov et al., 1977
[³H]-Spiperone	Neuroleptics	Creese and Synder, 1977
[³H]-Quinuclidinyl benzolate	Antidepressants	Innis et al., 1979
[³H]-Dihydroalprenolol	beta-Adrenergics	Innis et al., 1978
[³H]-Neurotensin	Neurotensin	Kitabgi et al., 1977
[³H]-Bradykinin	Bradykinin	Innis et al., 1981
[³H]-Substance P	Substance P	Nakata et al., 1978
[³H]-Somatostatin	Somatostatin	Aguilera and Parker, 1982
[³H]-Diazepam	Benzodiazepines	Mohler and Okada, 1977
[³H]-Kainic Acid	?	London and Coyle, 1979
[³H]-Nitrendipine	Calcium channel antagonists	Gould et al., 1983

[a]Listed are some established competitive binding assays currently used as RRAs or of potential interest in this regard. The compounds potentially detected by these methods are shown.

tory amino acid antagonist [³H]-2-amino-7-phosphonoheptanoic acid, for example, has recently been shown to label a population of binding sites in brain, and binding is inhibited by L-glutamate, but not L-aspartate (Ferkany and Coyle, 1984). Modification of this assay to quantitatively assess the former compound would not be difficult. Even using established competitive binding assays, additional RRAs for amino acids and related compounds could be developed. Thus, the specific binding of [³H]-strychnine to spinal cord membranes could easily be utilized to detect and quantitate endogenous glycine, whereas the [³H]L-glutamate binding assay might form the basis for an RRA to explore the concentration and distribution of N-acetylaspartyl-L-glutamate in brain.

The more likely use for RRAs involving amino acids arises from their utility to detect hitherto unidentified neurotransmitter-like substances in the brain. Using these techniques, compounds that influence glutamatergic (Zaczed et al., 1983; Ferkany et al., 1984) and GABAergic (Toffano et al., 1978; Guidotti et al., 1983) receptors, as well as receptors labeled by the excitatory amino acid analog [^{3}H]-kainic acid (Ferkany et al., 1984), have already been identified. Additional compounds of pharmacological and therapeutic importance, including the benzodiazepines, barbiturates, and the neurotoxin quinolinic acid, appear to interact with specific amino acid receptors in brain, and the search for possible endogenous ligands for these receptors is proceeding. Undoubtably, as more synthetic ligands for amino acid receptors become routinely available, the utility of RRAs to detect the additional substances active at these receptors will gain increasing importance.

8. Summary

Competitive binding assays are characterized by several properties, the most important of which are saturability and specificity. Beyond being useful to define the biochemical and pharmacologic aspects of neurotransmitter receptors, these techniques have been adapted to detect and quantify a variety of substances in tissues and fluids. When applied in this manner, ligand binding assays are referred to as radioreceptor assays (RRAs). Aside from their utility to measure known compounds or agents, RRAs are gaining increasing importance in the search for novel neurotransmitter-like substances in brain.

References

Aguilera G. and Parker D. S. (1982) Pituitary somatostantin receptors. *J. Biol. Chem.* **257**,1134–1137.
Baba A., Okumura S., Mizuo H., and Iwata H. (1983) Inhibition by diazepam and gamma-aminobutyric acid of depolarization-induced release of (^{14}C) cysteine sulfinate and (^{3}H) glutamate in rat hippocampal slices. *J. Neurochem.* **40**,280–284.
Baudry M., Smith E., and Lynch G. (1981a) Influence of temperature, detergents and enzymes on glutamate receptor binding and its regulation by calcium in rat hippocampal membranes. *Mol. Pharmacol.* **20**,280–286.
Baudry M. and Lynch G. (1981b) Characterization of two (^{3}H)glutamate

binding sites in rat hippocampal membranes. *J. Neurochem.* **36**,811–820.

Baudry M., Simon R., Smith E., and Lynch G. (1983) Regulation by calcium ions of glutamate receptor binding in hippocampal slices. *Eur. J. Pharmacol.* **90**,161–168.

Bennet J. P. (1979) Methods in binding studies, in Neurotansmitter Receptor Binding (Yamamura, H., Enna, S. J. and Kuhar, M. J. eds.) pp. 57–90. Raven Press, New York.

Bernasconi R., Bittiger H., Heid J., and Martin P. (1980) Determination of GABA levels by a (^{3}H)muscimol radioreceptor assay. *J. Neurochem.* **34**,614–618.

Braestrup C. and Squires, R. F. (1977) Benzodiazepine receptors in rat brain characterized by high affinity (^{3}H)diazepam binding. *Proc. Nat. Acad. Sci. (USA)* **9**,3805–3809.

Brehm L., Krogsgaard-Larsen P., and Jacobsen P. (1979) GABA-uptake inhibitors and structurally related pro-drugs, in GABA-Neurotransmitters (Krogsgaard-Larsen, P., Scheel-Kruger, J., and Kofod, H., eds.), pp. 247–262, Aademic Press, New York.

Burt D., Enna S. J., Creese I., and Snyder S. H. (1975) Dopamine receptor binding in the corpus striatum of mammalian brain. *Proc. Nat. Acad. Sci. (USA)* **72**,4655–4659.

Burt D. R. (1980) Basic receptor methods. II. Problems in interpretation in binding studies, in Receptor Binding Techniques, pp. 53–69. Society of Neuroscience, Bethesda, Md.

Butcher S. P., Collins J. F., and Roberts P. J. (1983) Characterization of the binding of D,L-[^{3}H]2-amino-4-phosphonobutyr ate to L-glutamate-sensitive binding sites on rat brain synaptic membranes. *Brit. J. Pharmacol.* **80**,355–364.

Calil H. M., Avery D. H., Hollister L. E., Creese I., and Snyder S. H. (1979) Serum levels of neuroleptics measured by dopamine radioreceptor assay and some clinical observations. *Psychiatr. Res.* **1**,39–44.

Chapman A. G., Collins J. F., Meldrum B. S., and Westerberg E. (1983) Uptake of a novel anticonvulsant 2-amino-7-phosphono (4,5-^{3}H) heptanoic acid into mouse brain. *Neurosci. Lett.* **37**,75–80.

Childers S. R., Simantov R., and Snyder S. H. (1977) Enkephalin: radioimmunoassay and radioreceptor assay in morphine-dependent rats. *Eur. J. Pharmacol.* **46**,289–293.

Coyle J. T. (1982) The clinical use of antipsychotic medication. Sympos. on Clin. Pharm. of Sympt. Control, *Med. Clin. North Am.* **66**,993–1009.

Creese I. and Snyder S. H. (1977) A simple and sensitive radioreceptor assay for antischizophrenic drugs in blood. *Nature* **270**,180–182.

Cuatrecasas P. and Hollenberg M. D. (1976) Membrane receptors and hormone action. *Adv. Prot. Chem.* **30**,251–451.

Enna S. J. and Snyder S. H. (1976) A simple, sensitive and specific radioreceptor assay for endogenous GABA in brain tissue. *J. Neurochem.* **26**,221–224.

Enna S. J. and Snyder S. H. (1977a) Gamma-aminobutyric acid (GABA) in human cerebrospinal fluid: radioreceptor assay. *J. Neurochem.* **28**,1121–1124.

Enna S. J. and Snyder S. H. (1977b) Influence of enyzmes, ions and detergents on GABA receptor binding in synaptic membranes of rat brain. *Mol. Pharmacol.* **13**,422–453.

Enna S. J. (1978a) Amino acid neurotransmitter candidates, in Annual Reports in Medicinal Chemistry (Humber, J., ed.) pp.41–50. Academic Press, New York.

Enna S. J. (1978b) Radioreceptor assay techniques for neurotransmitters and drugs, in Neurotransmitter Receptor Binding (Yamamura H., Enna S. J., and Kuhar M. J. eds.) pp. 127–139. Raven Press, New York.

Enna S. J., Ferkany J. W., and Krogsgaard-Larsen P. (1979) Pharmacological characterization of GABA receptors in different brain regions, in GABA-Neurotransitters (Krogsgaard-Larsen P., Scheel-Kruger J., and Kofod H., eds.) pp. 191-200. Academic Press, New York.

Enna S. J. (1980a) Radioreceptor assays, in Physico-chemical Methodologies in Psychiatric Research (Hanin I., and Koslow S. H., eds.) pp. 83–101. Raven Press, New York.

Enna S. J. (1980b) Basic receptor methods. I. in Receptor Binding Techniques, pp. 33–52. Society of Neuroscience, Bethesda, Md.

Enna S. J. and Gallagher J. P. (1983) Biochemical and electrophysiological characteristics of mammalian GABA receptors. *Int. Rev. Neurobiol.* **24**,181–212.

Fagg G. E., Foster A. C., Mena E. E., and Cotman C. W. (1982) Chloride and calcium ions reveal a pharmacologically distinct population of L-glutamate binding sites in synaptic membranes: correspondence between biochemical and electrophysiological data. *J. Neurosci.* **2**,958–965.

Fagg G. E., Mena E. E., Monoghan D. T., and Cotman C. W. (1983) Freezing eliminates a specific population of L-glutamate receptors in synaptic membranes. *Neurosci. Lett.* **38**,157–162.

Ferkany J. W., Smith L. A., Seifert W. E., Caprioli R. M., and Enna S. J. (1978) Measurement of gamma-aminobutyric acid (GABA) in blood. *Life Sci.* **22**,2121–2128.

Ferkany J., Zaczek R., Marckl A., and Coyle J. T. (1984) Glutamate-containing dipeptides enhance specific binding at glutamate receptors and inhibit specific binding at kainate receptors in rat brain. *Neurosci. Lett.* **44**,281–286.

Ferkany J. and Coyle J. T. (1983) Specific binding of [^{3}H]-2-amino-7-phosphonoheptanoic acid to rat brain membranes in vitro. *Life Sci.* **33**,1295–1305.

Ferkany J. W. and Coyle J. T. (in press) Receptors for excitatory amino acids and excitatory amino acid-like compounds in the mammalian central nervous system, in *Neuromethods: Neurochemistry IV— Receptor Binding* (Boulton A. A., Baker G. B., and Hrdina P., eds.) Humana Press, Inc., Clifton, New Jersey.

Fields J. Z., Reisine T. D., and Yamamura H. I. (1977) Biochemical demonstration of dopaminergic receptors in rat and human brain using (^{3}H)spiroperidol. *Brain Res.* **136**,578–584.

Foster A. and Fagg G. (1984) Acidic amino acid binding in mammalian neuronal membranes: their characteristics and relationship to synaptic receptors. *Brain. Res. Rev.* **7**,103–184.

Garcin F. and Coyle J. T. (1976) Ontogenetic development of (^{3}H)naloxone binding sites and endogenous morphine-like factors in rat brain, in Opiates and Endogenous Opioid Peptides (Kostelitz, H. W., ed.) pp. 41–48. Elsevier/North Holland, Amsterdam.

Gould R. J., Murphy K. M., and Snyder S. H. (1983) A simple sensitive radioreceptor assay for calcium antagonist drugs. *Life Sci.* **33**,2665–2672.

Guidotti A., Konkel D. R., Ebstein B., Corda M. G., Wise B. C., Krutzsch H., Meek J. L., and Costa E. (1982) Isolation, characterization and purification to homogeneity of a rat brain protein (GABA-modulin). *Proc. Nat. Acad. Sci. (USA)*, **79**, 6084–6088.

Hughes J., Smith J. W., Kosterlitz H. W., Fothergill L. A., Morgan B. A., and Morris A. R. (1975) Identification of two related pentapeptides from brain with potent opiate agonist activity. *Nature*, **258**, 577–578.

Innis R. B., Bylund D. B., and Snyder S. H. (1978) A simple, sensitive and specific radioreceptor assay for beta-adrenergic antagonist drugs. *Life Sci.* **23**,2031–2038.

Innis R. B., Tune L., Roek R., DePaulo R., U'Prichard D., and Snyder S. H. (1979) Tricyclic antidepressant radioreceptor assay. *Eur. J. Pharmacol.* **58**,473–477.

Innis R. B., Manning D. C., Stewart J. M., and Snyder S. H. (1981) (^{3}H)Bradykinen receptor binding in mammalian tissue membranes. *Proc. Nat. Acad. Sci.(USA)*, **78**,2630–2634.

Iwata H., Yama gami S., Mizuo H., and Baba A. (1982) Cysteine sulfinic acid in the central nervous system: uptake and release of cysteine sulfinic acid by a rat brain preparation. *J. Neurochem.* **38**, 1268–1274.

Kitabgi P., Carraway R., Retschofen J. V., Granier C., Morgat J. L., Menez A., Leeman S., and Freychet P. (1977) Neurotensin: specific binding to synaptic membranes from rat brain. *Proc. Nat. Acad. Sci. (USA)*, **74**,1846–1850.

Leeb-Lundberg F., Napios C., and Olsen R. W. (1981) Dihydropicrotoxinin binding sites in mammalian brain: interaction with convulsant and depressant benzodiazepines. *Brain Res.* **216**,399–408.

Lefkowitz F., Roth J., and Pastan I. (1970) Radioreceptor assay of adrenocorticotropin hormone: new approach to assay of polypeptide hormones in plasma. *Science,* **170**,633–635.

Leysen, J. E., Nemegeers C. J., Tollenaere J. P., and Laduron P. M. (1978) Serotonergic component of neuroleptic receptors. *Nature,* **272**,168–171.

List, S. J. and Seeman P. (1981) Resolution of dopamine and serotonin component of (^{3}H)spiperone binding to rat brain regions. *Proc. Nat. Acad. Sci. (USA)*, **78**,2620–2624.

London E.D. and Coyle J. T. (1979) Specific binding of (^{3}H)kainic acid to receptor sites in rat brain. *Mol Pharmacol.* **15**,492–505.

Luthin G. R. and Wolfe B. B. (1984) Comparison of [^{3}H]pirenzepine and [^{3}H]quinuclidinyl benzilate binding to muscarinic cholinergic receptors in rat brain. *J. Pharmacol. Exp. Therap.* **228**,648–655.

Massotti M., Guidotti H., and Costa E. (1981) Characterization of benzodiazepine and gamma-aminobutyric acid recognition sites and their endogenous modulators. *J. Neurosci.* **1**,409–418.

Mohler H. and Okada T. (1977) Benzodiazepine receptor: demonstration in the central nervous system. *Science*, **198**,849–851.

Monaghan P., McMillis M. D., Chamberlin A. R., and Cotman C. W. (1983) Synthesis of [^{3}H]-2-amino-4-phosphonobutyric acid and characterization of its binding to rat brain membranes: a selective ligand for the chloride/calcium-dependent class of L-glutamate binding sites. *Brain. Res.* **278**,137–144.

Moody J. W., Pert C. B., Rivier J., and Brown M. R. (1978) Bombesin: specific binding to rat brain membranes. *Proc. Nat. Acad. Sci. (USA)*, **75**,5372–5376.

Nakata Y., Kusaka Y., Segawa T., Yajima H., and Kitagawa K. (1978) Substance P: regional distribution and specific binding to synaptic membranes in rabbit central nervous system. *Life Sci.* **22**,259–268.

Olsen R. W. and Leeb-Lundberg F. (1981) Convulsant and anticonvulsant drug binding sites related to the GABA receptor/ionophore system, in Neurotransmitters, Seizures and Epilepsy (Morselli P., Lloyd K., Loscher W., Meldrum B., Chir B. and Reynolds E., eds.) pp151–164. Raven Press, New York.

Pasternak G. W., Goodman R., and Snyder S. H. (1975) An endogenous morphinelike factor in mammalian brain. *Life Sci.* **16**,1756–1770.

Peroutka S. J. and Snyder S. H. (1983) Multiple serotonin receptors and their physiological significance. *Fed. Proc.* **42**,213–217.

Pitman R. N., Minnina K. P., and Molinoff P. B. (1980) Ontogeny of beta$_1$ and beta$_2$ adrenergic receptors in rat cerebellum and cerebral cortex. *Brain Res.* **188**,357–368.

Recasens M., Varga V., Nanopoulos D., Saadova F., Vincendon G., and Benavides J. (1982) Evidence for cysteine sulfinate as a neurotransmitter. *Brain Res.* **239**,153–173.

Scatchard G. (1949) The attraction of proteins for small molecules and ions. *Ann. N. Y. Acad. Sci.* **51**,660–672.

Simantov R. and Snyder S. H. (1976a) Isolation and structure identification of a morphine-like peptide "enkephalin" in bovine brain. *Life Sci.* **18**,781–788.

Simantov R. and Snyder S. H. (1976b) Morphine-like peptides, leucine enkephalin and methionine enkephalin: interactions with the opiate receptor. *Mol. Pharmacol.* **12**,987–998.

Simantov R. and Snyder S. H. (1976c) Brain pituitary mechanisms: pituitary opiate receptor binding, radioimmunoassays for methionine enkephalin and leucine enkephalin and (^{3}H)enkephalin interactions with the opiate receptor, in Opiates and Endogenous Opioid Peptides (Kostelitz H. W., ed.) pp. 41-48. Elsevier/North Holland, Amsterdam.

Simantov R., Childers S. R., and Snyder S. H. (1977) Opioid peptides: differentiation by radioimmunoassay and radioreceptor assay. *Brain Res.* **135**,358–367.

Slevin J. T. and Coyle J. T. (1981) Ontogeny of receptor binding sites for (^{3}H)glutamic acid and (^{3}H)kainic acid in the rat cerebellum. *J. Neurochem.* **37**,531–533.

Snyder S. H., Chang K. J., Kuhar M. J., and Yamamura H. I. (1975) Biochemical identification of the mammalian muscarinic cholinergic receptor. *Fed. Proc.* **34**,1915–1921.

Snyder S. H. (1975) The glycine synaptic receptor in the mammalian central nervous system. *Br. J. Pharmacol.*, **53**,473–484.

Tallman J. F., Thomas J. W., and Gallager D. W. (1978) GABAergic modulation of benzodiazepine binding site sensitivity. *Nature,* **274,** 383–388.

Terenius L. and Wahlstrom A. (1974) Inhibitors of narcotic receptor binding in brain extracts and cerebrospinal fluid. *Acta Pharmacol. Toxicol.* **35**,(S1)55.

Terenius L. and Wahlstrom A. (1975) Search for an endogenous ligand for the opiate receptor. *Acta Physiol. Scand.* **94**,74.

Teschenaker H., Opheim K. E., Cox B. M., and Goldstein R. (1975) A peptide-like substance from pituitary that acts like morphine. *Life Sci.* **16**, 1771–1776.

Toffano G., Guidotti A., and Costa E. (1978) Purification of an endogenous protein inhibitor of the high affinity binding of gamma-aminobutyric acid to synaptic membranes of rat brain. *Proc. Nat. Acad. Sci. (USA),* **75**,4024–4028.

Toggenburger G., Felix D., Cuenod M., and Henke H. (1982) In vitro release of endogenous beta-alanine, GABA and glutamate and electrophysiological effect of beta-alanine in pigeon optic tectum. *J. Neurochem.* **39**,176–183.

Tune L. E., Creese I., DePaulo R., Slavney P. R., Coyle J. T., and Snyder S. H. (1980) Clinical state and serum neuroleptic levels measured by radioreceptor assay in schizophrenia. *Am. J. Psychiatr.* **137**,187–190.

Watkins J. C. and Evans R. H. (1981) Excitatory amino acid transmitters. *Ann. Rev. Pharmacol. Toxicol.* **21**,165–204.

Williams J. D. (1982) Cholecystokinin: a hormone and a neurotransmitter. *Biomedical Res.* **3**, 107–121.

Yammaura H., Enna S. F., and Kuhar M. J., eds. (1978) *Neurotransmitter Receptor Binding.* Raven Press, New York.

Young A. and Snyder S. H. (1973) Strychnine binding associated with glycine receptors of the central nervous system. *Proc. Nat. Acad. Sci. USA,* **70**,2832–2836.

Zaczek R., Koller K., Cotter R., Heller D., and Coyle J. T. (1983) N-acetylaspartyl glutamate: an endogenous peptide with high affinity for a brain "glutamate" receptor. *Proc. Nat. Acad. Sci. USA,* **80,**1116–1119.

Chapter 7

Immunocytochemical Techniques

JANG-YEN WU AND CHIN-TARNG LIN

1. Introduction

One of the most powerful tools in the identification of various neuronal types and their connectivities is immunocytochemical localization of specific neuronal markers, e.g., synthetic enzymes for neurotransmitters at precise cellular and subcellular levels. Several methods exist for the localization of tissue antigens, at both the light and electron microscopic (EM) levels. The fluorescein-labeled antibody or immunofluorescent method developed by Coons and his coworkers (1958) for the localization of tissue antigens with the light microscope has been employed for the localization of enzymes involved in the metabolism of neurotransmitters such as dopamine-β-hydroxylase (Hartman, et al., 1972; Hartman, 1973), tyrosine hydroxylase (Pickel et al., 1975a), phenylethanolamine-N-methyltransferase (Hökfelt et al., 1974), DOPA decarboxylase (Hökfelt et al., 1973), and L-glutamate decarboxylase (GAD) (Kataoka et al., 1984). There are some deficiencies in this otherwise sensitive and specific method, however, such as the masking of the specific fluorescence by the inherent background fluorescence of tissue, and a lack of permanence of the preparations. Nakane and Pierce (1966, 1967) reported that enzymes of small molecular weight, such as acid phosphatase or peroxidase, could be conjugated to antibodies by bifunctional reagents. The enzyme-labeled antibodies were enzymatically and immunologically active and were reacted with tissues that then were stained histochemically for the enzyme with deposition of reaction product at the antigenic site. Unlike the preparations by the fluorescent antibody technique, enzyme-labeled antibody preparations are permanent and can be observed with an ordinary light microscope. In addition to its appli-

155

cation in light microscopy, the enzyme-labeled antibody technique has also been applied to the ultrastructural localization of tissue antigens by EM because of the electron opacity of the stained reaction product (Nakane and Pierce, 1967). This technique has been applied to the localization of the enzymes involved in the metabolism of neurotransmitters, e.g., tyrosine hydroxylase (Pickel et al., 1975b, 1977), tryptophan hydroxylase (Joh et al., 1975) and GAD (Saito et al., 1974a; McLaughlin et al., 1974, 1975a,b). One difficulty involved in the peroxidase-labeled method is to obtain an enzymatically and immunologically active peroxidase–antibody conjugate with a molar ratio of 1:1. This problem can be minimized by the unlabeled antibody–enzyme method as described by Sternberger et al. (1970) and Sternberger (1974). This technique is based on the interaction of tissue antigen with its specific antiserum (primary antiserum), followed by incubation with an antiserum to the IgG in the primary antiserum (anti-IgG), followed by the addition of peroxidase and purified antibody to peroxidase (raised in the same animal species as the primary antiserum). The latter reacts with the anti-IgG already present in the tissue section and, therefore, immobilizes the peroxidase at the desired site. The unlabeled antibody–enzyme method has several advantages over labeled antibody methods in that it avoids nonspecificity resulting from the chemical labeling process itself, loss of antibody activity due to labeling, and interference of unlabeled with labeled antibodies. Petrali et al. (1974) reported that a peroxidase–antiperoxidase (PAP) method was 4–5 times more sensitive than the sequential application of antiperoxidase and peroxidase in the unlabeled antibody–enzyme method. Since the peroxidase method has been widely used in immunocytochemical studies of various nervous antigens, the authors will limit the scope of the review to the methodology and application of this technique. Furthermore, the authors also would like to use the GABA system as a model to illustrate the approach that has been taken in their laboratory, from the purification of antigens, production of monoclonal and polyclonal antibodies, to the immunocytochemical localization of GABAergic marker GAD at light and EM levels.

2. Preparation of Antigens

2.1. Antigens for Monoclonal Antibodies

Although the hybridoma technique does not require pure antigen for the production of specific monoclonal antibodies, it still requires a specific probe such as pure antigen for the detection of antibody-producing clones. Furthermore, the purer the antigen used in the initial immunization, the easier the screening step will be in the detection of the specific antibody-producing clones. Hence, the antigens used in the authors' laboratory for the initial immunization are usually between 10–50% pure, and those used in the screening step are the homogeneous preparations.

2.2. Antigens for Polyclonal Antibodies

A homogeneous antigen preparation is a prerequisite for obtaining a monospecific polyclonal antibody, which in turn is a prerequisite for obtaining any meaningful immunocytochemical results. In the last few years, we have purified GAD (Wu et al., 1973; Wu, 1976, 1983) and GABA-transaminase (GABA-T) (Schousboe et al., 1973; Wu, 1976) from mouse brain, GAD from catfish brain (Su et al., 1979; Wu, 1983), GAD and cysteinesulfinic/cysteic acids decarboxylase (CSAD/CAD) from bovine brain (Wu, 1982), choline acetyl transferase (CAT) from the electric organ of *Torpedo* (Brandon and Wu, 1978), catfish brain (Su et al., 1980) and bovine brain (Wu et al., 1979), and clathrin from bovine brain (Garbern and Wu, 1981). More recently, GAD has been purified to homogeneity from rat brain (Denner and Wu, 1984). The first step of purification is to determine regional and subcellular distributions of these enzymes, so that some degree of purification can be obtained by choosing proper regions and fractions as the enzyme source. Briefly, the methods include the initial extraction of GAD from crude mitochondrial fraction, followed by ammonium sulfate fractionations and column chromatographies on Sephadex G-200, calcium phosphate gel, and DEAE-Sephadex, and finally preparative gel electrophoresis. The purity of the antigen was established from the following physical and chemical criteria.

2.2.1. Polyacrylamide Gel Electrophoresis

The purified enzyme (25–70 μg) revealed as a single protein band with the location of the enzyme activity corresponding to the location of the protein band on polyacrylamide gel electrophoresis, suggesting that the enzyme preparations are homogeneous in terms of size and charge (Wu, 1982, 1983).

2.2.2. Sedimentation Equilibrium Analysis

The purified mouse brain GAD and GABA-T preparations appeared to be homogeneous in size under various conditions as judged from the linear plot of the logarithm of concentrations (c) against the squares of the distances (r) from the center of rotation of points of interest in high-speed sedimentation equilibrium runs in both H_2O and D_2O solutions. Furthermore, GAD preparations treated with dissociating reagents, e.g., guanidine HCl and 2-mercaptoethanol, also appeared to be monodisperse in the high-speed sedimentation equilibrium runs (Wu, 1976).

2.2.3. Gradient Polyacrylamide Gel Electrophoresis

On 3.6–25% or 6–10% gradient polyacrylamide gel electrophoresis, the purified GAD from catfish or rat brain migrated as a single protein band with comigrating enzyme activity (Su et al., 1979; Wu, 1983; Denner and Wu, 1984).

2.2.4. Sodium Dodecyl Sulfate (SDS)–Polyacrylamide Gel Electrophoresis

The purified clathrin migrated as a single protein band on SDS–polyacrylamide gel electrophoresis. Furthermore, a single, symmetric peak was obtained when the gel was traced with a densitometer (Garbern and Wu, 1981).

2.2.5. Isoelectric Focusing

The purified rat brain GAD also migrated as a single protein band that corresponded to the enzyme activity on narrow range isoelectric focusing gels (Denner and Wu, 1984).

2.2.6. Immunodiffusion

In the immunodiffusion test, antibodies against partially purified GAD also formed a sharp, single band with the purified GAD preparations, whereas the crude preparations showed multiple bands (Matsuda et al., 1973), suggesting that the purified GAD preparation is immunochemically homogeneous.

3. Preparation of Antibodies

3.1. Production of Monoclonal Antibodies

Recently, we have employed the hybridoma technique originally developed by Köhler and Milstein (1975, 1976) for the production of monoclonal antibody against GAD and CAT (Wu et al., 1982b) and aspartate amino transferase (AAT) (Lin and Chen, 1982; Lin et al., 1983a). Briefly, the hybridoma was prepared according to the method of Kennett et al. (1978) and Kennett (1979) with some modifications. Each of two rats (or mice) were immunized with purified mouse (or rat) brain GAD or CAT intraperitoneally for 6 wk. The sera from the immunized rats (or mice) had been shown to contain a high titer of antibody against GAD or CAT. The spleens were removed from the immunized rat or mice and perfused with culture medium at several sites, thereby forcing the spleen cells into the culture medium. The erythrocytes were lysed with NH_4Cl, and the spleen cells were fused with plasmacytoma cell line (P3 × 63AG8) with a ratio of 10:1 in 50% polyethylene glycol for 5 min.

The hybrids were evenly suspended and gently distributed into 10 microplates (1 drop/well). The next day, an additional drop of the 2 × HAT medium (hypoxanthine, thymidine, and aminopterin) was added. The wells were fed two additional drops of HT medium (without aminopterin) 6–7 d later. Clones appeared 17 d later.

The positive wells were identified by screening the supernatant liquid for production of the antibody against GAD or CAT by ELISA test (enzyme-linked immunosorbent assay) using peroxidase-labeled goat anti-rat (or mouse) IgG as second antibody and purified GAD or CAT as antigen. These antibody-producing hybrids were then subcloned by limiting dilution, rescreened with the pure antigen, and grown in 250 mL bottles. Clones yielding positive responses in an additional ELISA were then injected into the peritoneal cavity of immunosuppressed mice. Ascites fluid was collected every 3–4 d, retested in the ELISA, and stored at −20°C.

3.2. Production of Polyclonal Antibodies

Once the purity of antigens had been established by vigorous physical and chemical tests as described above, various amounts (usually between 1.5 and 30 μg) of pure antigen were emulsified in complete Freund's adjuvant and injected biweekly into rabbit at subscapular muscles. Animals were bled after the fifth injection, which was followed by intermittent monthly boosters of 10–30 μg. Serum was isolated by brief centrifugation after clotting for 1 h at 37°C and 8–16 h at 4°C. This technique has been successfully used in our laboratory for the production of antibodies against various proteins purified from the nervous system. For instance, a total of 7–150 μg purified GABA-T (Saito et al., 1974b) and GAD from mouse brain (Saito et al., 1974c; Wu, 1983), CAT from the electric organ of *Torpedo* (Brandon and Wu, 1977), neurofilament protein from *Myxicola* (Lasek and Wu, 1976), GAD, CAT, and CSAD from bovine brain (Wu et al., 1979; Wu, 1982), GAD and CAT from catfish (Su et al., 1980, 1983; Wu, 1983), and GAD from rat brain (Denner et al., 1984) were able to evoke production of specific antibodies in rabbits.

An alternative method is to inject the gel slice that contains the protein of choice into the rabbit. This is particularly useful in those cases where trace impurities are copurified with the protein of choice, but they can be clearly separated by the gel electrophoresis system. We have used this technique to prepare antibody against SDS–polyacrylamide gel electrophoresis-purified clathrin, and have shown that the antibody is specific to clathrin (Garbern and Wu, 1981). If it is desirable, the γ-globulin fraction can be obtained from serum by precipitation with ammonium sulfate at 50% saturation. IgG fraction can be prepared by DE cellulose chromatography, as previously described (Wu, 1983).

3.3. Characterization of Antibodies

3.3.1. Immunodiffusion and Immunoelectrophoresis Tests

Immunodiffusion gels were made of 1% agarose in 15 mM sodium phosphate, 140 mM sodium chloride, pH 7.4 (PBS) containing 0.02% sodium azide, 1 mM 2-aminoethylisothiouronium bromide (AET), and 0.2 mM pyridoxal phosphate (PLP), pH 7.4. Serum or concentrated antigen (1–10 μL) was placed in the wells. Plates were kept covered in a humidified atmosphere at 4°C for 24–72 h. Immunoelectrophoresis was performed in gels containing 1% agarose, 25 mM Tris, 192 mM glycine, 1 mM AET, 0.2 mM PLP, and 0.01% sodium azide, pH 8.4. Samples of less than 5 μL were applied and electrophoresis performed for 1.5 h at 4°C with 10 V/cm constant voltage. The electrophoresis running buffer was made of 25 mM Tris, 192 mM glycine, 1 mM AET, 0.2 mM PLP, and 0.5% 2-mercaptoethanol, pH 8.4. One lane was cut from the gel and stained for protein, and a parallel lane, 1 cm wide and approximately 1 mm thick, was cut in 0.5 cm lengths and assayed for GAD activity. Slices were macerated in disposable culture tubes containing 125 μL of 200 mM potassium phosphate, 1 mM AET, 0.2 mM PLP, and 1 mM EDTA, pH 6.0. After adding 12.5 μL L-[1-^{14}C]glutamic acid (5 μCi/mL, 40 mM sodium glutamate), GAD activity was measured. Narrow troughs (approximately 1 mm wide, 10 cm long) were then cut between the remaining lanes in the direction of electrophoresis. Preimmune or immune serum (100 μL) was placed in the troughs and immunodiffusion carried out for 12–36 h.

In both immunodiffusion and immunoelectrophoresis tests, when antiserum against the purified GAD was tested with a crude GAD preparation, a single precipitation band was obtained (Wong et al., 1974). Furthermore, this precipitation band contained GAD activity, suggesting that the precipitation band is GAD–anti–GAD complex and the antiserum is specific to GAD (Wu, 1983; Denner et al., 1984). In order to increase the sensitivity of immunodiffusion and immunoelectrophoresis, the immunodiffusion plate or immunoelectrophoresis gel after an extensive wash can be stained for protein with Coomassie blue or by an autoradiographic technique using ^{125}I-labeled protein A because of the affinity of protein A for the Fc (constant fragment) portion of the mammalian IgG. We have used this technique to visualize the precipitation band that may be too faint to be photographed, as in the case of clathrin and anticlathrin (Garbern and Wu, 1981). We have also used the peroxidase-labeled antibody method which was introduced by Nakane and Pierce (1966, 1967) for localizing tissue antigens to stain the precipitation bands formed in agar immunodiffusion tests (Matsuda et al., 1973). Briefly, crude GAD preparations were incubated with anti-GAD serum; this was followed by extensive washing and incubation with peroxidase-labeled goat anti-rabbit IgG. Finally, the agar plate was stained with 0.01% (v/v) H_2O_2 and 2.5 mM 3,3'-diamino-benzidine in 0.05M Tris-HCl, pH 7.6.

3.3.2. Immunoprecipitation and Enzyme Inhibition

The enzyme activity of mouse brain GAD was inhibited to a maximum of about 50% by incubating with excess of anti-GAD IgG for 24 h at 4°C (Saito et al., 1974c). Almost all of the enzyme activity was precipitated, presumably in the form of GAD–anti-GAD complex, when GAD (17 µg protein) was incubated with an approximately equal amount of anti-GAD serum (Matsuda et al., 1973). For rat brain GAD, Gad–anti-GAD complexes were precipitated using the *Staphylococcus aureus* Cowan type I (SAC) procedure as described (Denner et al., 1984). Briefly, 80 µL (2 mg/mL) of partially purified GAD preparation was incubated for 4–16 h at 4°C with an equal volume of immune serum or IgG fractions at various dilutions. Unless stated otherwise, all dilutions were made with, and pellets resuspended in, PBS containing 1 mM AET, 0.2 mM PLP, 1 mM EDTA, pH 7.0. SAC was prepared daily by two washes in PBS containing 0.05% Tween-20, followed by incubation at room temperature for 30 min with agitation, and finally washing three more times in PBS. Sac was resuspended in the original volume and 15 µL added to each tube. Samples were incubated at room temperature with agitation for 30 min and then briefly centrifuged. Supernatants and pellets were then assayed for GAD activity. The following controls were performed: no immune serum; no SAC; preimmune serum instead of immune serum; or neither immune serum nor SAC. Total GAD activity in the supernatant and pellet in the presence of immune serum or anti-GAD IgG compared to those with the preimmune serum or control IgG indicated the degree of inhibition of enzyme activity by the antiserum. When the amount of GAD and SAC was kept constant, the increase of antiserum decreased GAD activity in the supernatant with a concomitant increase in GAD activity in the pellet. If IgG fraction was used, a complete and quantitative immunoprecipitation was achieved with the SAC procedure. GAD activity was only slightly inhibited by anti-GAD IgG. No effect was observed with preimmune IgG.

3.3.3. Microcomplement Fixation Test

Since serum from unimmunized rabbits also interfered with the fixation of complement, it was necessary to employ IgG for microcomplement fixation tests. In a typical experiment, 6 μg of anti-GAD IgG and 5–100 ng of GAD was used in each tube. Isotonic veronal buffer containing Ca^{2+} and Mg^{2+} and 0.1% gelatin was used as a diluent (Mayer, 1961). Sensitization of sheep blood cells with an optimal amount of hemolysin was performed as described (Mayer, 1961).

Titration of complement and the microcomplement fixation tests were performed according to the method of Wasserman and Levine (1961) using anti-GAD IgG in a final volume of 1.4 mL. Hemolysis was allowed to proceed for 60 min. After the removal of precipitate, the optical density of the supernatant fluid was determined at 413 nm spectrophotometrically. For mouse brain GAD, 50% fixation of complement was obtained with about 6 μg of anti-GAD IgG and 40 ng of GAD. Furthermore, the fixation curves obtained with the partially purified GAD and the purified GAD preparations became superimposable when the amount of GAD protein was estimated from the specific activities of GAD preparations, suggesting that the antiserum is specific to GAD only (Saito et al., 1974c). Microcomplement fixation can be employed to determine the actual quantity of antigen protein in a crude preparation. For instance, in case of GAD, the extent of complement fixed was roughly proportional to the amount of GAD in the range of 15–50 ng of GAD. In addition, microcomplement fixation tests can be used to distinguish antigens with subtle differences in their structure. For instance, microcomplement fixation has been reported to be capable of distinguishing lactate dehydrogenases with a single amino acid difference (Wilson et al., 1964). GAD from various species can also be distinguished by a microfixation test. The complement fixation curves obtained with crude GAD preparations from mouse, rat, and human were very similar. The maximal degree of fixation with GAD from calf, rabbit, and guinea pig was about 40–65%. GAD from quail, pigeon, frog, and trout did not react at all under these conditions (Saito et al., 1974c; Wu, 1983).

3.3.4. Immunodot Analysis

Immunodot analysis was performed according to the procedure described (Denner et al., 1984). Nitrocellulose discs were made with a paper hole punch to fit snugly into the bottom of standard 96-well microtiter plates. Unless otherwise mentioned, all procedures were performed at room temperature with constant gentle agitation. GAD preparations of different degrees of purity diluted in PBS were spotted on the discs in 2 μL volumes containing variable amounts of protein. All subsequent dilutions and washes were with PBS containing 0.05% Tween-20, 0.1% bovine serum albumin, pH 7.4 (buffer B). After incubation at 37°C for 4 h, discs were washed three times for 5 min each. Preimmune serum or antiserum was diluted 1:100, added in 100 μL aliquots per well, and incubated for 2 h. After three washes, 100 μL of peroxidase conjugated goat anti-rabbit IgG diluted 1:200 was added and incubated for an additional 1 h. Discs were then washed three additional times. The substrate (0.05% 3,3'-diaminobenzidine tetrahydrochloride in 50 mM Tris, pH 7.6) was passed through a 0.22 μm filter, made to 0.01% hydrogen peroxide, and used within 5 min. The reaction was terminated after 2 min by aspiration of the substrate followed by several rapid washes. For rat brain GAD, 5–500 ng of pure GAD or 50–5000 ng of 10% pure GAD preparation were spotted on nitrocellulose discs. The intensity of the peroxidase reaction product was roughly proportional to the amount of antigen spotted. In addition, the intensity of the reaction product was fairly similar for discs containing approximately equal amounts of GAD protein, but different amounts of total protein. The specificity of this reaction was further indicated in the controls where no reaction product was observed. The controls contained 500 and 50 ng of pure GAD protein treated with preimmune serum instead of immune serum or without the second antibody (Denner et al., 1984).

3.3.5. Enzyme-Linked Immunosorbent Assay

Aliquots of 50 μL of GAD solution containing 50–1000 ng of GAD were added to microtiter wells and incubated with constant agitation at room temperature overnight. Wells were washed three times with 200 μL of buffer B. Fifty μL of either diluted (1:100) immune or preimmune serum, culture medium, or Ascites fluid was added and incubated for 2 h. The wells were washed three times. After additional incubation with peroxidase-conjugated goat anti-rabbit IgG (for polyconal antibodies from rabbits) or peroxidase-conjugated rabbit anti-mouse IgG (for monoclonal antibodies from mice) diluted 1:200 with buffer B, wells were washed three more times and treated with 100 μL of substrate [0.05% 2,2'-azinodi(3-ethylbenzthiazoline) sulfonic acid, 0.1M citric acid, pH 4.2, 0.02% hydrogen peroxide] for 15 min at room temperature. The reaction was stopped by adding 0.001% sodium azide. The intensity of reaction product is proportional to the amount of antigen added and is independent of the purity of the preparation since a similar pattern was obtained either with pure or partially purified GAD preparations. The specificity of the antibodies was further indicated by the lack of reaction product when GAD was treated with preimmune serum or control culture medium (Wu et al., 1982a; Denner et al., 1984).

4. Preparation of Tissues

4.1. Tissue Fixation

For immunocytochemical localization of certain amino acid transmitter enzymes in the central nervous system, a good preservation of tissue morphology and antigen antigenicity is essential. In general, perfusion fixation is better than immersion fixation. For perfusion fixation one may perfuse the fixative into the animal circulatory system through the left ventricle of the heart or through the abdominal aorta.

In the past, we have used cardiac perfusion to localize calmodulin in the rat cerebellum (Lin et al., 1980), clathrin in the rat cerebellum and kidney (Lin et al., 1982a), and CSAD in the rat cerebellum (Chan-Palay et al., 1982c). Similarly, we have used retrograde aortic perfusion to localize cytosolic AAT in rat organs (Lin and Chen, 1983), and GAD, GABA-T, CSAD, AAT, and somatostatin in the rat retina (Lin et al., 1983b). It seems that the aortic perfusion method is better than the cardiac perfusion because the former is easier to handle and the circulation can always be maintained throughout the whole procedure before the fixative is administered. The choice of fixative depends mainly on the nature of the antigen and the degree of preservation of tissue structure required. In order to determine the optimal condition of fixation, it is necessary to use different concentrations of paraformaldehyde and glutaraldehyde and to check the tissue morphology and the immunoreactivity. Most enzymes could be fixed in 4% paraformaldehyde and 0.1% glutaraldehyde mixture for achieving a reasonable preservation of morphology and antigenicity. The fixative can be prepared in 0.04M sodium phosphate buffer, pH 7.4, containing 0.02% $CaCl_2$. Once the animal has been perfused with 200–600 mL of fixative, the brain can be removed and the area of interest should be dissected out immediately and immersed overnight in a weaker fixative, such as 2% paraformaldehyde, in order to remove excess glutaraldehyde.

4.2. Tissue Sectioning

For tissue sectioning, one can use one of the following three methods: (1) paraffin block sectioning, (2) cryostat sectioning, and (3) vibratome sectioning. Since most amino acid transmitter enzyme markers in the brain are labile and sensitive to the paraffin embedding processes, this method is not recommended. However, this method is suitable for other antigens, such as apoVLDL-II, ovalbumin, and high density lipoproteins in chicken liver (Lin and Chan, 1980, 1981). Cryostat sectioning is perhaps the most gentle treatment for tissue antigenicity and is suitable for light microscopic immunohistochemical studies of transmitter enzymes. The tissue has to be protected with cryoprotectants, such as 20% sucrose or 10% dimethylsulfoxide (Lin et al., 1983b). Five-micrometer sections can be easily obtained. For EM immunocytochemistry, cryostat sectioning is not suitable since the ice crystal formation during tissue freezing always gives an inferior ultrastructure. Vibratome sectioning is easier to handle, although it is difficult to obtain a section thinner than 20 μm. Usually, a 50-μm section can be obtained without difficulty. Although somewhat too thick for immunohistochemistry, it is quite good for EM immunocytochemistry. For both light and EM localization in one tissue section, vibratome sectioning is the best choice (Lin, 1980; Lin et al., 1980; Lin et al., 1982a,b; Lin and Chan, 1982; Lin et al., 1983).

5. Immunocytochemical Procedures

5.1. Light Microscopy

An indirect immunocytochemical method using PAP complex has been extensively used for antigen localization at the light microscopic level. The tissue sections that are obtained either from the cryostat mounted on glass slides or the vibratome sectioning floated free in buffer solution are treated with normal serum (or IgG) from the animal from which the second antibody was produced. For example, if the first antibody is prepared in the rabbit, the normal serum should be taken from an animal such as the goat, which produces the antiserum against rabbit IgG. After 10 min incubation, the sections are then incubated with the first antibody overnight at room temperature. The concentration of the

antiserum used is determined by its titer. Usually, a dilution of
1:40–1:200 in phosphate buffer should be tried. After incubation
with the first antibody, the sections are washed twice with phosphate buffer, 10 min each, followed by incubation with second
antibody (for example, goat antiserum against rabbit IgG) with
1:40 dilution (or 0.1 mg/mL) for 1 h at room temperature. Sections
are then incubated for 15 min at room temperature with PAP complex diluted 1:100. Sometimes the optimal concentration of first
and second antibody and PAP has to be determined by serial dilutions, such as 1:40, 1:80, 1:100, and 1:200. After twice washing
with buffer, 10 min each, the sections are then incubated for 5 min
at room temperature with the peroxidase substrate 3,3'-
diaminobenzidine-4HCl (DAB). DAB solution is prepared as
0.05% in 0.05M Tris-HCl, pH 7.6. Before use, the DAB solution is
added to 0.01% H_2O_2. The reaction is terminated by removal of
the substrate. Sections are washed in buffer and mounted with
50% glycerine in distilled water. A representative result obtained
by this procedure is demonstrated in Fig. 1–3.

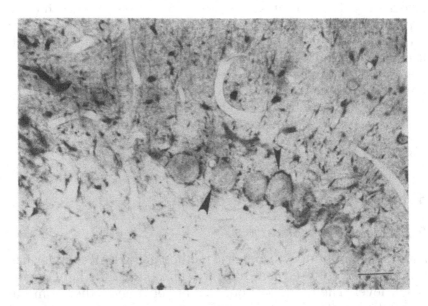

Fig. 1.   Immunohistochemical localization of GAD in rat cerebellum. A rat perfused through the aorta with 600 mL of 4%
paraformaldehyde and 0.1% glutaraldehyde mixture in 0.04M phosphate buffer, pH 7.4, containing 0.02% $CaCl_2$. A 50-μm section was obtained with a vibratome. The section was stained with rabbit antiserum
against mouse GAD using the PAP method as described in the text.
Punctate reaction product is shown surrounding Purkinje cells (arrowheads). Some neuronal dendrites and stellate cell bodies in the molecular layer are also stained. Bar = 10 μm.

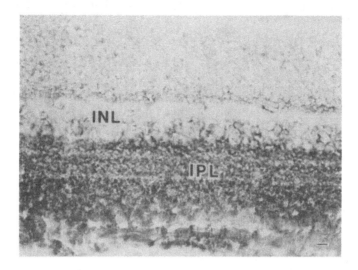

Fig. 2. Immunohistochemical localization of GAD in rat retina. Fixation conditions are similar to those in Fig. 1. A 12-μm section was obtained from the cryostat. GAD reaction product is localized in certain neuronal cell bodies in the inner nuclear layer (INL) as well as in many processes within the inner plexiform layer (IPL). The multistratified laminae of GAD staining is clearly discernible in the IPL. Bar = 10 μm.

5.2. Electron Microscopy

An indirect immunoperoxidase method using peroxidase-labeled second antibody seems better for EM immunocytochemical localization than the PAP method as described for the light microscopic immunohistochemistry. The advantage of the PAP method is its sensitivity through amplification of the peroxidase molecules. This high degree of amplification of peroxidase reaction product is actually disadvantageous for EM immunocytochemistry because the excess of peroxidase reaction product tends to reduce the degree of resolution of fine ultrastructures. Since EM localization requires a good preservation of morphology, the fixative and the washing buffer should always include 8.5% of sucrose in $0.1M$ sodium phosphate buffer in order to increase the osmolarity in the solution. For tissue staining, the 50 μm sections obtained from the vibratome sectioning are incubated with normal serum as described before. After 10 min, sections are transferred and incubated with the first antibody for 1 h at room temperature. Sections are then washed, and further incubated with peroxidase-labeled second antibody for 1 h. After extensive washing, the sections are incubated with peroxidase substrates as described. Sections are refixed in 2.5% glutaraldehyde for 30 min, washed in

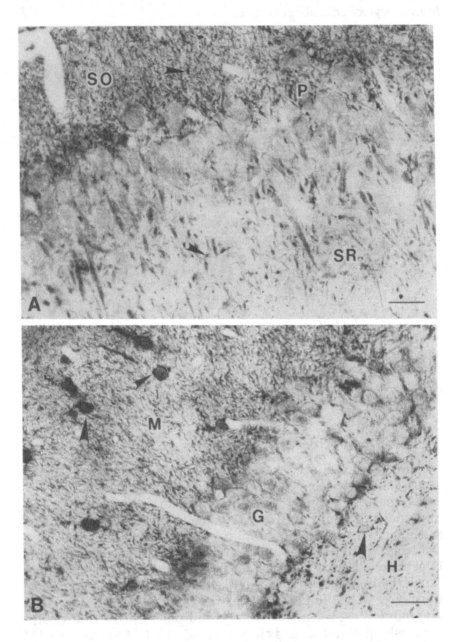

Fig. 3. Immunohistochemical localization of CSAD in rat hippo-campus. Fixation and tissue sectioning are similar to Fig. 1. The sections were stained with rabbit antiserum against CSAD. A: This picture, taken from hippocampus area CA3, shows punctate reaction product around the pyramidal cells (P). Some reaction product (arrowheads) is also demonstrated in the stratum oriens (SO) and stratum radiatum (SR). (× 1020). B: This picture, taken from dentate gyrus, shows reaction product in the cell bodies of interneurons (arrows) in the molecular layer (M) and hilar region (H). The neuronal processes in both regions and around granule cell bodies (G) are also stained. Bar = 10 μm.

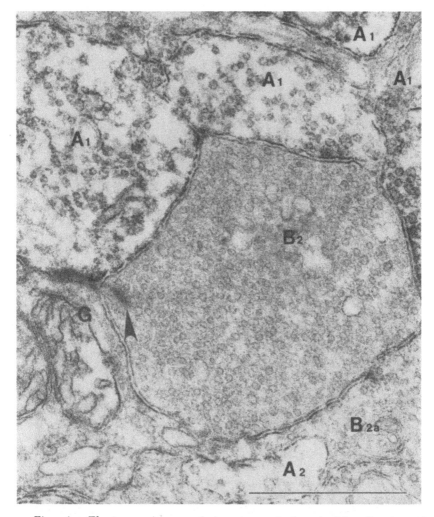

Fig. 4. Electron microscopic immunocytochemical localization of GAD in the inner plexiform layer of rat retina. A rat was perfused with 600 mL of 4% paraformaldehyde and 0.2% glutaraldehyde mixture in $0.1M$ phosphate buffer, pH 7.4, containing 8.5% sucrose and 0.01% $CaCl_2$. A 50-μm section obtained from the vibratome sectioning was stained with antiserum against GAD. An unstained bipolar terminal (B_2) (at the center) makes synapses to three stained amacrine (A_1) terminals, one unstained possible bipolar terminal (B_{2a}), and one unstained ganglion cell dendrite (G). A typical bipolar dyad containing the synaptic ribbon (arrowhead) and its postsynaptic, stained, amacrine terminal (A_1), and unstained ganglion cell dendrite (G) is clearly discernible. Those three stained amacrine terminals (A_1) also make direct contact to each other although there are no visible synapses. Other stained (A_1) and unstained (A_2) amacrine processes are also noted. Reaction product in each stained terminal is seen to be associated mainly with synaptic vesicles and synaptic membrane (counterstained with lead citrate). ($\times$ 34,000). Bar = 1 μm.

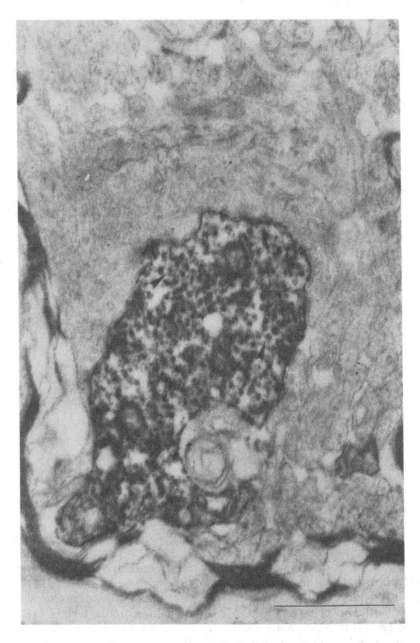

Fig. 5. Electron microscopic immunocytochemical localization of CSAD in rat substantia nigra. The fixation and tissue sectioning were similar to Fig. 4. The section was stained with antiserum against CSAD. One of the neuronal terminals is stained. Reaction product is seen to be associated with the terminal plasma membrane and some organelles. In addition, most synaptic vesicles in this terminal contain reaction product. The membranes of some synaptic vesicles (arrowheads) are clearly discernible. ($\times$ 33,000). Bar = 1 μm.

buffer, postfixed in 1% OsO_4 for 1 h, washed in distilled water, prestained in saturated uranyl acetate in 50% ethanol for 10 min, dehydrated with ethanol, and finally embedded in Spurr's medium for thin sectioning. All thin sections are observed under EM without further staining. A representative EM picture using this technique is shown in Figs. 4 and 5.

5.3. Control Experiment

Both light and EM immunostainings require vigorous control. For some unknown reason(s), the control section may sometimes show false positive staining. Whenever such false staining appears in the control section, the result from the specific antibody staining should not be counted. Several control experiments should be included to verify the specificity of the staining. (1) Use the preimmune serum to replace the specific antiserum. (2) Use specific antiserum, which has been previously absorbed with excess pure antigen. (3) Use antibodies against antigens that are known to be absent in that particular tissue or area. (4) Omit the first antibody. (5) Use peroxidase substrate only.

6. Concluding Remarks

The immunochemical and immunocytochemical techniques described in this review have been proven to be quite powerful in the identification and elucidation of various transmitter systems, particularly the GABAergic neurons and processes, in the vertebrate system. These include rabbit retina (Brandon et al., 1979, 1980; Wu et al., 1981), goldfish and frog retina (Lam et al., 1979; Brandon et al., 1980; Wu et al., 1981; Zucker et al., 1984), monkey striate cortex and geniculate complex (Hendrickson et al., 1981, 1983), cat hippocampus and visual cortex (Somogy et al., 1983a,b), rat cerebellum (Saito et al., 1974a; McLaughlin et al., 1974, 1975b; Chan-Palay et al., 1979), habenula (Gottesfeld et al., 1980, 1981), hypothalamus and pituitary gland (Vincent et al., 1982), pancreas (Vincent et al., 1983), dentate gyrus (Goldwitz et al., 1982; Kosaka et al., 1984), spinal cord (McLaughlin et al., 1975a; Hunt et al., 1981), dorsal lateral geniculate nucleus (O'Hara et al., 1983), septum (Panula et al., 1984), and human cortex (Vincent et al., 1984). Other systems such as those involving taurine (Wu, 1982; Chan-Palay et al., 1982c,d; Lin et al., 1983b, 1984), acetylcholine (Chan-Palay et al., 1982a,b), peptides (e.g., motilin (Chan-Palay et al., 1981) and somatostatin (Lin et al., 1983c), and coated vesicle protein, clathrin (Lin et al., 1982a), have also been successfully identified using similar approaches.

References

Brandon C. and Wu J.-Y. (1977) Electrophoretic and immunochemical characterization of choline acetyltransferase from *Torpedo. Soc. Neurosci. Abs.* **3,** 404.

Brandon C. and Wu J.-Y. (1978) Purification and properties of choline acetyltransferase from *Torpedo. J. Neurochem.* **30,** 791–797.

Brandon C., Lam K. M. K., and Wu J.-Y. (1979) The γ-aminobutyric acid system in rabbit retina: Localization by immunocytochemistry and autoradiography. *Proc. Natl. Acad. Sci. USA* **76,** 3557–3561.

Brandon C., Lam D. M. K., Su Y. Y. T., and Wu J.-Y. (1980) Immunocytochemical localization of GABA neurons in the rabbit and frog retina. *Brain Res. Bull.* **5** (Suppl. 2), 21–29.

Chan-Palay V., Palay S. L. and Wu J.-Y. (1979) Gamma-aminobutyric acid pathways in the cerebellum studied by retrograde and antero-grade transport of glutamic acid decarboxylase antibody after in vivo injections. *Anat. & Embryol.* **157,** 1–14.

Chan-Palay V., Nilaver G., Palay S. L., Beinfeld M. C., Zimmerman E. E., Wu J.-Y. and O'Donohue T. L. (1981) Chemical heterogeneity in cerebellar purkinje cells: Existence and coexistence of glutamic acid decarboxylase-like and motilin-like immunoreactivities. *Proc. Natl. Acad. Sci. USA* **78,** 7787–7791.

Chan-Palay V., Engel A. G., Palay S. L. and Wu J.-Y. (1982a) Synthesizing enzymes for four neuroactive substances in motor neurons and neuromuscular junctions: Light and electron micro-scopic immunocytochemistry. *Proc. Natl. Acad. Sci. USA* **79,** 6717–6721.

Chan-Palay V., Engel A. G., Wu J.-Y. , and Palay S. L. (1982b) Coexist-ence in human and primate neuromuscular junctions of enzymes synthesizing acetylcholine, catecholamine, taurine, and γ-amino-butyric acid. *Proc. Natl. Acad. Sci. USA* **79,** 7027–7030.

Chan-Palay V., Lin C. T., Palay S., Yamamoto M. and Wu J.-Y. (1982c) Taurine in the mammalian cerebellum: Demonstration by autoradi-ography with [^{3}H]taurine and immunocytochemistry with antibod-ies against the taurine-synthesizing enzyme, cysteine-sulfinic acid decarboxylase. *Proc. Natl. Acad. Sci. USA* **79,** 2695–2699.

Chan-Palay V., Palay S. L., Li C., and Wu J.-Y. (1982d) Sagittal cerebel-lar micro-bands of taurine neurons: Immunocytochemical demon-stration by using antibodies against the taurine-synthesizing en-zyme cysteine sulfinic acid decarboxylase. *Proc. Natl. Acad. Sci. USA* **79,** 4221–4225.

Coons A. H. (1958) Fluorescent Antibody Methods, in *General Cyto-chemical Methods* (Danielli, J. F., ed.) p. 399. Academic Press, N.Y.

Denner L. A. and Wu J.-Y. (1985) Purification and characterization of L-glutamate decarboxylase from whole rat brain. *J. Neurochem.* (In Press).

Denner L. A., Lin C.-T., Song G.-X., and Wu J.-Y. (1985) Production and characterization of polyclonal and monoclonal antibodies to rat brain L-glutamate decarboxylase *Brain Res.*, (In press)

Garbern J.-Y. and Wu J.-Y. (1981) Purification and characterization of clathrin from bovine brain. *J. Neurochem.* **36**, 602–612.

Goldwitz D., Vincent S. R., Wu J.-Y., and Hökfelt T. (1982) Immunohistochemical demonstration of plasticity in GABA neurons of the adult rat dentate gyrus. *Brain Res.* **238**, 413–420.

Gottesfeld Z., Brandon C., Jacobowitz D. M., and Wu J.-Y. (1980) The GABA system in the mammalian habenula. *Brain Res. Bull.* **5**, (Suppl. 2), 1–6.

Gottesfeld Z., Brandon C., and Wu J.-Y. (1981) Immunocytochemistry of glutamate decarboxylase in the deafferented habenula. *Brain Res.* **208, 181–186.**

Hartman B.K. (1973) Immunofluorescence of dopamine-β-hydroxylase. *J. Histochem. Cytochem.* **21,** 312–332.

Hartman B. K., Zide D., and Udenfriend S. (1972) The use of dopamine-β-hydroxylase as a marker for the central noradrenergic nervous system in rat brain. *Proc. Natl. Acad. Sci. USA* **69, 2722–2726.**

Hendrickson A. E., Hunt S., and Wu J.-Y. (1981) Immunocytochemical localization of glutamic acid decarboxylase in monkey striate cortex. *Nature* **292,** 605–607.

Hendrickson A. E., Ogren M. P., Vaughn J. E., Barber R. P., and Wu J.-Y. (1983) Light and electron microscopic immunocytochemical localization of glutamic acid decarboxylase in monkey geniculate complex: Evidence for GABAergic neurons and synapses. *J. Neuroscience* **3,** 1245–1262.

Hökfelt T., Fuxe K., and Goldstein M. (1973) Immunohistochemical localization of aromatic L-amino acid decarboxylase (DOPA decarboxylase) in central dopamine and 5-hydroxytryptamine nerve cell bodies of the rat. *Brain Res.* **53, 175–180.**

Hökfelt T., Fuxe K., Goldstein M., and Johansson O. (1974) Immunohistochemical evidence for the existence of adrenaline neurons in the rat brain. *Brain Res.* **66,** 235–251.

Huang B. H. and Wu J.-Y. (1984) Ultrastructural studies on catecholaminergic terminals and GABAergic neurons in nucleus tractus solitarii of the medulla oblongata of rat. *Brain Res.* 302, 57–67.

Hunt S. P., Kelly J. S., Emson P. C., Kimmel J. R., Miller R. J., and Wu J.-Y. (1981) An immunohistochemical study of neuronal populations containing neuropeptides or GABA within the superficial layers of the rat dorsal horn. *Neuroscience* **6,** 1883–1898.

Joh T. H., Shikimi T., Pickel V. M., and Reis D. J. (1975) Brain tryptophan hydroxylase: Purification of, production of antibodies to, and cellular and ultrastructural localization in serotonergic neurons of rat midbrain. *Proc. Natl. Acad. Sci. USA* **72,** 3575–3579.

Kataoka Y., Gutman Y., Guidotti A., Panula P., Wroblewski Y., Cosenza-Murphy D., Wu J.-Y., and Costa E. (1984) The intrinsic GABAergic system of adrenal chromaffin cells. *Proc. Natl. Acad. Sci. USA,* 81, 3218–3222.

Kennett R. H. (1979) Cell fusion, in *Methods in Enzymology* (Jakoby W. and Pastan J., eds.) pp. 345–349. Academic Press, New York.

Kennett R. H., Denis J., Tung A., and Klinman N. (1978) Hybrid plasmacytoma production: Fusions with adult spleen cells, monoclonal spleen fragments, neonatal spleen cells and human spleen cells. *Curr. Top. Microbiol. Immunol.* **81,** 77–91.

Köhler G. and Milstein C. (1975) Continuous cultures of fused cells secreting antibody of predefined specificity. *Nature* **256,** 495–497.

Köhler G. and Milstein C. (1976) Derivation of specific antibody-producing tissue culture and tumor lines by cell fusion. *Eur. J. Immunol.* **6,** 514–519.

Kosaka T., Hama K. and Wu J.-Y. (1984) GABAergic synaptic boutons in the rat dentate gyrus. *Brain Res.* 293, 353–359.

Lam D. M. K., Su Y. Y. T., Swain L., Marc R. E., Brandon C., and Wu J.-Y. (1979) Immunocytochemical localization of glutamic acid decarboxylase in goldfish retina. *Nature* **278,** 565–567.

Lasek R. J. and Wu J.-Y. (1976) Immunochemical analysis of the proteins comprising myxicola (10nm) neurofilaments. *Soc. Neurosci. Abstr.* **2,** 40.

Lin C. T. (1980) Immunoelectron microscopic localization of immunoglobulin G in human placenta. *J. Histochem. Cytochem.* **28,** 339–346.

Lin C. T. and Chan L. (1980) Effects of estrogen on specific protein synthesis in the cockerel liver: An immunocytochemical study on major apoproteins in very low density and high density lipoproteins and albumin. *Endocrinology* **107, 70–75.**

Lin C. T. and Chan L. (1981) Estrogen regulation of yolk and non-yolk protein synthesis in the avian liver: An immunocytochemical study. *Differentiation* **18, 105–114.**

Lin C. T. and Chan L. (1982a) Localization of apoVLDL-II, a major apoprotein in very low density lipoproteins in the estrogen-treated cockerel liver by immunoelectron microscopy. *Histochemistry* **76,** 237–246.

Lin C. T. and Chen L. (1982b) Comparison of polyclonal and monoclonal antibodies for immunocytochemical localization of cytosolic aspartate aminotransferase alpha subform in rat liver. *J. Histochem. Cytochem.* **30,** 578 (Abstr.)

Lin C. T. and Chen L. H. (1983) Production and characterization of an antibody to cytosolic aspartate aminotransferase and immunolocalization of the enzyme in rat organs. *Lab. Invest.* **48,** 718–725.

Lin C. T., Dedman J. R., Brinkley B. R., and Means A. R. (1980) Localization of calmodulin in rat cerebellum by immunoelectron microscopy. *J. Cell Biol.* **85,** 473–480.

Lin C. T., Garbern J. and Wu J.-Y. (1982a) Light and electron microscopic immunocytochemical localization of clathrin in rat cerebellum and kidney. *J. Histochem. Cytochem.* **30,** 853–863.

Lin C. T., Mukai K., and Lee C. Y. (1982b) Electron microscopic immunocytochemical studies of hCG binding and endocytosis in rat ovary. *Cell Tissue Res.* **224,** 647–653.

Lin C. T., Chen L. H., and Chan T. S. (1983a) A comparative study of polyclonal and monoclonal antibodies for immunocytochemical localization of cytosolic aspartate aminotransferase in rat liver. *J. Histochem. Cytochem.* **31,** 920–927.

Lin C. T., Li H. Z., and Wu J.-Y. (1983b) Immunocytochemical studies and comparison of regional distribution of L-glutamate decarboxylase, gamma aminobutyric acid transaminase, cysteinsulfinic acid decarboxylase, aspartate aminotransferase and somatostatin in rat retina. *Brain Res.* **270,** 273–283.

Lin C. T., Su C. C., Palmer W., and Chan L. (1983c) Localization of somatostatin in dog pancreas by immunoelectron microscopy. *Tissue and Cell* **15,** 259–270.

Lin C. T., Song G.-X., Li H.-Z., and Wu J.-Y. (1985) Ultrastructural demonstration of L-glutamate decarboxylase and cysteine sulfinic acid decarboxylase in rat retina by immunocytochemistry. *Brain Res.* (In Press).

Matsuda T., Wu J.-Y., and Roberts E. (1973) Immunochemical studies on glutamic acid decarboxylase from mouse brain. *J. Neurochem.* **21,** 159–166.

Mayer M. M. (1961) Complement and Complement Fixation, in *Experimental Immunochemistry* (Kabat E. A. and Mayer M. M., eds.), Thomas, Springfield, IL.

McLaughlin B. J., Wood J. G., Saito K., Barber R., Vaughn J. E., Roberts E., and Wu J.-Y. (1974) The fine structural localization of glutamate decarboxylase in synaptic terminals of rodent cerebellum. *Brain Res.* **76,** 377–391.

McLaughlin B. J., Barber R., Saito K., Roberts E., and Wu J.-Y. (1975a) Immunocytochemical localization of glutamate decarboxylase in rat spinal cord. *J. Comp. Neurol.* **164,** 305–322.

McLaughlin B. J., Wood J. G., Saito K., Roberts E., and Wu J.-Y. (1975b) The fine structural localization of glutamate decarboxylase in developing axonal processes and presynaptic terminals of rodent cerebellum. *Brain Res.* **85,** 355–371.

Nakane P. K. and Pierce G. B., Jr. (1966) Enzyme-labeled antibodies: Preparation and application for the localization of antigens. *J. Histochem. Cytochem.* **14,** 929–931.

Nakane P. K. and Pierce G. B., Jr. (1967) Enzyme-labeled antibodies for the light and electron microscopic localization of tissue antigens. *J. Cell Biol.* **33,** 307–318.

O'Hara P. T., Lieberman A. R., Hunt S. P. and Wu J.-Y. (1983) Neural elements containing glutamic acid decarboxylase (GAD) in the dorsal lateral geniculate nucleus of the rat: Immunohistochemical studies by light and electron microscopy. *Neuroscience* **8,** 189–211.

Panula P., Revuelta A. V., Cheney D. L., Wu J.-Y., and Costa E. (1984) An immunohistochemical study on the location of GABAergic neurons in rat septum. *J. Comp. Neurol.* **222,** 69–80.

Petrali P., Hinton M., Moriarty C., and Sternberger A. (1974) The unlabeled antibody enzyme method of immunocytochemistry. Quantitative comparison of sensitivities with and without peroxidase–antiperoxidase complex. *J. Histochem. Cytochem.* **22,** 782–801.

Pickel V. M., Joh T. H., Field P. M., Becker C. G., and Reis D. J. (1975a) Cellular localization of tyrosine hydroxylase by immunohistochemistry. *J. Histochem. Cytochem.* **23,** 1–12.

Pickel V. M., Joh T. H., and Reis D. J. (1975b) Ultrastructural localization of tyrosine hydroxylase in noradrenergic neurons of brain *Proc. Natl. Acad. Sci. USA* **72,** 659–663.

Pickel V. M., Joh T. H., and Reis D. J. (1977) Light and electron microscopic localization of tyrosine hydroxylase by immunocytochemistry, in *Structure and Function of Monoamine Enzymes* (Usdin E., Weiner N., and Youdin M. B. H., eds.) p. 821–833. Marcel Dekker, Inc., New York.

Saito K., Barber R., Wu J.-Y., Matsuda T., Roberts E. and Vaughn J. E. (1974a) Immunohistochemical localization of glutamic acid decarboxylase in rat cerebellum. *Proc. Natl. Acad. Sci. USA* **71,** 269–273.

Saito K., Schousboe A., Wu J.-Y., and Roberts E. (1974b) Some immunochemical properties and species specificity of GABA-α-ketoglutarate transaminase from mouse brain. *Brain Res.* **65,** 287–296.

Saito K., Wu J.-Y., and Roberts E., (1974c) Immunochemical comparisons of vertebrate glutamate acid decarboxylase. *Brain Res.* **65,** 277–285.

Schousboe A., Wu J.-Y., and Roberts E. (1973) Purification and characterization of the 4-aminobutyrate-2-ketoglutarate transaminase from mouse brain. *Biochemistry* **12,** 2868–2873.

Somogyi P., Freund T., Wu J.-Y., and Smith A. D. (1983a) The secretion of Golgi impregnation procedure. II. Immunocytochemical demonstration of glutamate decarboxylase in Golgi-impregnated neurons and in their afferent and efferent synaptic boutons in the visual cortex of the cat. *Neuroscience* **9,** 475–490.

Somogyi P., Smith A. D., Nunzi M. G., Gorio A., Takagi H., and Wu J.-Y. (1983b) Glutamate decarboxylase immunoreactive neurons and distribution of their synaptic terminals on pyramidal neurons in the hippocampus of the cat, with special reference to the axon initial segment. *J. Neuroscience* **3,** 1450–1468.

Sternberger L. A. (1974) *Immunocytochemistry,* p. 129. Prentice-Hall, Inc., Englewood Cliffs, NJ.

Sternberger L. A., Hardy P. H., Jr., Cuculis J. J., and Meyer H. G. (1970) The unlabeled antibody enzyme method of immunohistochemistry. Preparation and properties of soluble antigen–antibody complex (horseradish peroxidase-anti-horseradish peroxidase) and its use in identification of spirochetes. *J. Histochem. Cytochem.* **18,** 315–333.

Su Y. Y. T., Wu J.-Y., and Lam D. M. K. (1979) Purification of L-glutamic acid decarboxylase from catfish brain. *J. Neurochem.* **33,** 169–179.

Su Y. Y. T., Wu J.-Y., and Lam D. M. K. (1980) Purification and some properties of choline acetyltransferase from catfish brain. *J. Neurochem.* **34,** 438–445.

Su Y. Y. T., Wu J.-Y., and Lam D. M. K. (1983) Species specificaties of L-glutamic acid decarboxylase: Immunochemical comparisons. *Neurochem. Intl.* **5,** 587–592.

Vincent S. R., Hökfelt T., and Wu J.-Y. (1982) GABA neuron systems in hypothalamus and the pituitary gland: Immunohistochemical demonstration using antibodies against glutamate decarboxylase. *Neuroendocrinology* **34,** 117–125.

Vincent S. R., Meyerson B., Sachs C., Hökfelt T., Goldstein M., Wu J.-Y., Brown M., Elde R. P., Terenius L., Steinbusch H. W. M., Vernhofstad A. A. A., Culleo A. C., Fahrenkrug J., Rehfeld J., Dockray G. J., and Kimmel J. (1984) Neurotransmitters in the human cortex: An immunohistochemical study. *Neuroscience,* in press.

Vincent S. R., Hökfelt T., Wu J.-Y., Elde R. P., Morgan L. M., and Kimmel J. R. (1983) Immunohistochemical studies of the GABA system in the pancreas. *Neuroendocrinology* **36,** 197–204.

Wasserman E. and Levine L. (1961) Quantitative microcomplement fixation and its use in the study of antigenic structure by specific antigen–antibody inhibition. *J. Immunol.* **87,** 290–295.

Wilson A. C., Kaplan N. O., Levine L., Pesce A., Reichlin M. and Allison W. S. (1964) Evolution of lactic dehydrogenases. *Fed. Proc.* **23,** 1258–1266.

Wong E., Schousboe A., Saito K., Wu J.-Y., and Roberts E. (1974) Glutamate decarboxylase and GABA-transaminase from six mouse strains. *Brain Res.* **68,** 133–139.

Wu J.-Y. (1976) Purification and properties of L-glutamate decarboxylase (GAD) and GABA-aminotransferase (GABA-T), in *GABA in Nervous System Function* (Roberts E., Chase T. and Tower D., eds.), pp. 7–55. Raven Press, New York, N.Y.

Wu J.-Y. (1982) Purification and characterization of cysteic/cysteine sulfinic acids decarboxylase and L-glutamate decarboxylase in bovine brain. *Proc. Natl. Acad. Sci. USA* **79,** 4270–4274.

Wu J.-Y. (1983) Preparation of glutamic acid decarboxylase as immunogen for immunocytochemistry, in *Neuroimmunocytochemistry (IBRO Handbook Series: Methods in the Neurosciences)* (Cuello A. C., ed.), pp. 159–191. John Wiley & Sons, Ltd., Sussex.

Wu J.-Y., Matsuda T., and Roberts E. (1973) Purification and characterization of glutamate decarboxylase from mouse brain. *J. Biol. Chem.* **248,** 3029–3034.

Wu J.-Y., Su Y. Y. T., Brandon C., Lam D. M. K., Chen M. S., and Huang W. M. (1979) Purification and immunochemical studies of GABA-, acetyl- choline-, and taurine-synthesizing enzymes from bovine and fish brains. *Proc. 7th Meet. Int. Soc. Neurochem.,* p. 662.

Wu J.-Y., Brandon C., Su Y. Y. T., and Lam D. M. K. (1981) Immunocytochemical and autoradiographic localization of GABA system in the vertebrate retina. *Mol. Cell. Biochem.* **39,** 229–238.

Wu J.-Y., Lin C.-T., Brandon C., Chan D.-S., Möhler H., and Richards J. G. (1982a) Regulation and immunocytochemical characterization of GAD, in *Cytochemical Methods in Neuroanatomy* (Palay S. and Palay V., eds.) pp. 279–296. Alan R. Liss, Inc.

Wu J.-Y., Lin C.-T., Denner L., Su Y. Y. T., and Chan D. S. (1982b) Monoclonal antibodies of GABA- and acetylcholine-synthesizing enzymes. *Proc. Meet. Amer. Soc. Neurochem.*, Abstract 13(1), 92.

Zucker C., Wu J.-Y., and Yazulla S. (1984) Non-correspondence of ^{3}H-GABA uptake and GAD localization: Two potential markers of GABAergic neurons. *Brain Res.* 298, 154–158.

Chapter 8

In Vitro Autoradiographic Localization of Amino Acid Receptors and Uptake Sites

W. SCOTT YOUNG, III

1. Introduction

Some amino acids may function as neurotransmitters and are presumably released at nerve terminals to interact with specific receptors postsynaptically. Biochemical studies of high-affinity uptake, a potential marker of nerve terminals and of receptor binding, have been employed successfully to elucidate the role of amino acids in the central nervous system (W. Walz and J.W. Ferkany, this volume). The recent use of autoradiographic techniques to study uptake and receptors has permitted much greater anatomical resolution than homogenate methods. Other advantages include the ability to study different conditions in consecutive sections, to study human tissues, and to circumvent the blood–brain barrier.

The principal goal of this chapter is to present a light microscopic method for receptor autoradiography and then review its application to amino acid receptors in tissue sections. Autoradiographic methods to study uptake sites in tissue sections and slices also will be reviewed.

2. In Vitro Light Microscopic Autoradiography

2.1. Rationale

The ability to study ligand binding to receptors has greatly expanded our appreciation and understanding of putative amino

179

acid transmitters and drugs that interact with their receptors. The elegant methods of receptor binding presume that the receptors have a high degree of selectivity under appropriate conditions, such that relatively high-affinity "specific" binding of, generally, a radioactively labeled ligand rises above the lower affinity "nonspecific" background. Several excellent works treat this subject in greater detail (Cuatrecasas and Hollenberg, 1976; Yamamura et al., 1984). The autoradiographic approach is based upon the application of binding techniques to tissue sections with subsequent apposition of radiosensitive emulsion to the sections (Young and Kuhar, 1979a, 1981). A given receptor can be studied in this fashion if it survives the freezing necessary to cut the thin sections. This can be ascertained by performing biochemical studies on the slide-mounted sections and comparing results with those obtained in in vitro homogenate studies. Appropriate studies might include those of kinetics (including the dissociation constant, K_d, and number of receptors, B_{max}); stereospecificity; pharmacological specificity; ion, nucleotide, and other drug effects; and regional studies. If these studies confirm the viability of the receptor (and nearly all do survive), one can proceed to autoradiography.

The in vitro autoradiographic approach to the study of receptors has several important advantages. The primary advantage is the resolution at the micron level, which also means that binding is being performed on nanogram tissue samples. This represents an increase in sensitivity over homogenate assays of 10^6–10^7 fold. Obviously, this also translates into much less costly assays. From a practical and theoretical point of view as well, the ratio of ligand to receptor concentrations remains constant during the incubations as a result of the very small quantities used. Autoradiography of amino acid receptors has not been possible after intravenous injections because of metabolism and the blood–brain barrier. These problems are readily avoided through in vitro autoradiography. As with homogenates, a controlled binding environment is possible. This technique enables one to study different receptors or conditions in consecutive, relatively undisrupted, tissue sections. And, finally, human postmortem tissue is amenable to in vitro autoradiography.

The potential limitations of diffusion of ligand from receptor and morphological defects induced by freezing or by processing of tissue and emulsion do not seem significant. Ultrastructural resolution necessitates ligands with irreversible or nearly irreversible binding. Although this has not been accomplished with amino acid receptors, the γ-aminobutyric acid (GABA)-influenced benzodiazepine receptor has been photoaffinity-labeled (Möhler et al., 1981). The general approach to ultrastructural studies will probably entail immunocytochemistry with antibodies against the receptors.

2.2. Method

The method described here (Fig. 1) is based on the one developed by Young and Kuhar (1979a) in which labeling is performed on a thaw-mounted tissue section, nonspecific binding is washed off, the section is rapidly dried, and radiosensitive emulsion is apposed. An alternative approach employing fixation of tissue prior to coating the sections with liquid emulsion (Herkenham and Pert, 1982) has not, to my knowledge, been applied to the study of amino acid receptors. As discussed by Kuhar and Unnerstall (1982), care must be taken in using this latter technique because significant loss and diffusion of ligand may occur.

2.2.1. Tissue Preparation

Tissue is removed and placed on a brass chuck with surrounding OCT compound (Lab-Tek Products, Naperville, IL) or brain paste and then quickly frozen in liquid nitrogen or on dry ice. The tissue may first be perfused in vivo by a light fixative, such as 0.1% formaldehyde and/or 5–15% sucrose to improve histological features if the receptor is not affected. After freezing, the tissue is stored at −70°C until used.

2.2.2. Receptor Labeling

Eight to sixteen micron sections are cut at −18°C on a cryostat/microtome and thaw-mounted onto acid-washed, subbed (dipped into a solution of 0.5 g gelatin and 50 mg chrome alum in 100 mL water) slides and allowed to air dry before being returned to −20°C for storage. This latter step allows the tissue to adhere to the slide so that the sections do not float off during subsequent steps.

The slides with tissue sections are brought to room temperature before incubation with the radiolabeled ligands in buffer at various temperatures (Table 1). After incubation, the sections are rinsed in buffer, briefly dipped in distilled water to remove salts, and rapidly dried under a stream of cool, dry air to minimize possible diffusion. In biochemical studies, this latter step is unnecessary because the sections are wiped off with a piece of filter paper and counted in a scintillation counter. As discussed above, these preliminary biochemical studies are imperative to assess the receptor and derive proper conditions (incubation times, wash times, and so on) for the autoradiography.

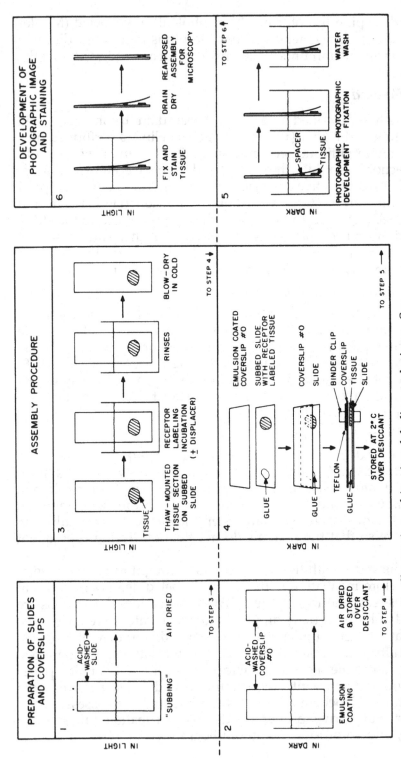

Fig. 1. Schematic illustration of in vitro labeling technique. See Section 2.2 for further details (from Young and Kuhar, 1979; with permission).

TABLE 1
Ligand Binding Conditions for In Vitro Autoradiography

Receptor	Ligand	Concentration	Incubation time and temperature	Rinse time (at ice-bath temperature)	Blank	Reference
GABA	^{3}H-muscimol	5 nM	40 min; ice-bath	1 min	200 μM GABA	Palacios et al., 1980
Glycine	^{3}H-strychnine	4 nM	20 min; ice-bath	5 min	1 mM glycine	Zarbin et al., 1981
Glutamate	^{3}H-glutamate	100 nM	30 min; 37°C	4 quick rinses	1 mM glutamate	Greenamyre et al., 1983
	^{3}H-kainate	15 nM	120 min; ice-bath	3 min	100 μM kainate	Unnerstall and Wamsley, 1983

2.2.3. Autoradiography

Once the receptors are appropriately labeled, either an emulsion-coated coverslip (Roth et al., 1974; Young and Kuhar, 1979a, 1981) or tritium-sensitive Ultrofilm (Palacios et al., 1981a; Penny et al., 1981; Rainbow et al., 1982b) is apposed to the tissue section. The former approach uses acid-washed glass coverslips (25 × 77 mm, Corning No. 0, Corning, NY) that have been dipped into Kodak NTB3 emulsion (Rochester, NY) diluted 1:1 with water at 43°C, air-dried for 3 h, and stored over desiccant for 24 h to 7 d. The emulsion-coated coverslips are attached to the slides with the tissue sections in the dark with glue (Super Glue No. 3, Loctite Corp., Cleveland, OH) that is placed on one end of the slide. A line is drawn with a grease pencil to prevent the glue from moving down the slide toward the tissue section. After the glue sets (about 30 s), squares of Teflon or cardboard (⅛ in. thick) are put on top of the coverslips, and the assemblies are held together with No. 20 binder clips. The assemblies are stored, desiccated in the dark, at 2–4°C for varying lengths of time. As a rule of thumb, we allow 6–8 wk of exposure for 1500 dpm/mg tissue and extrapolate from that value for other cases. Positive chemography, the spurious production of grains, may be assessed using buffer-exposed tissues. Negative chemography can be assessed by using emulsion-coated coverslips briefly exposed to light (Rogers, 1979).

After exposure, the binder clips are removed, and the coverslips gently bent away from the tissue sections with a round spacer. The emulsion is developed in Dektol (Kodak, 1:1 with water) for 2 min at 17°C, placed in Kodak Liquid Hardener (1:13 with water) for 15 s, fixed in Kodak Rapid Fix for 3 min, and rinsed in distilled water for 20 min. The tissues are then fixed in Carnoy's solution and stained for 30–60 s in Pyronin Y. An alternative fluorescent counterstain is ethidium bromide (2.5 × $10^{-7}M$ for 1 min; Sigma Chemical Co., St. Louis, MO) that can be viewed with a 546/590 nM exciter/barrier filter set. This counterstain avoids tissue stain background during darkfield observation. Before using Permount (Fisher, Fair Lawn, NJ) or D.P.X. (for ethidium bromide; BDH Chemicals, Poole, England) to mount the coverslips, the sections are dried for at least 1 h at 40°C to prevent fading of autoradiograms. These processing steps can be performed on many assemblies simultaneously through the use of various holders (Young and Kuhar, 1981). The autoradiograms are then examined microscopically under both brightfield (or fluorescent) and darkfield illumination. A calibrated eyepiece grid is used to count grains for quantitative studies (*see* below).

Tritium-sensitive Ultrofilm (LKB Industries, Rockville, MD) is exposed by placing the slides side-by-side against the film in an X-ray film cassette and the entire sheet is exposed at 4°C. After exposure, the film is developed for 5 min at 20°C in Kodak D19 developer, rinsed in a stop bath, and fixed in Kodak Rapid Fix for 5 min at 20°C. The film is rinsed in running water for 20 min at 20°C, rinsed in Photo-Flo solution, and hung to dry. Optical densities of these autoradiograms can be read by a variety of microdensitometers.

2.2.4. Further Issues

The tritium-sensitive film method provides a faster exposure medium at the expense of the decreased resolution afforded by the nuclear emulsion-coated coverslip method. And although the former method is easily adapted to computer analysis and presentation, the latter method maintains the registration of the autoradiogram over the tissue. Several laboratories have presented standardization techniques for quantitative analysis employing tissue (Rainbow et al., 1982b), brain paste (Unnerstall et al., 1982), or plastic (Penney et al., 1981; Lysz et al., 1982; Geary and Wooten, 1983) standards. Lysz et al. (1982) observed lack of covariation between standards of ^{14}C and ^{3}H and stressed the need to standardize using the ligand's isotope. This may reflect, in part, the ability of the higher-energy beta emissions of ^{14}C to reach deeper portions of the emulsion.

A further complication is the recent discovery that gray and white matter have different radioabsorbencies for tritium (Alexander et al., 1981). An approach to this problem soaks adjacent sections in a freely diffusable compound such as ^{3}H-isoleucine to permit appropriate compensation for differential absorbencies by different areas (Taylor et al., 1984). Further discussions of this and other issues are presented by Rogers (1979), Kuhar (1982, 1983, 1984), Wamsley and Palacios (1983), and Young and Kuhar (1979a, 1981).

3. Amino Acid Receptor Localization

3.1. γ-Aminobutyric Acid Receptors

Gamma-aminobutyric acid (GABA) is a major inhibitory neurotransmitter in the central nervous system of mammals, and its receptor was the first amino acid receptor visualized (Palacios et al., 1980). Also, this receptor has remained a popular one for autoradiographic studies (Table 2). The binding characteristics of GABA and the GABA receptor agonist, muscimol, have been reviewed in this volume by J.W. Ferkany. Muscimol has been the preferred ligand because of its weak affinity for the GABA uptake system and for GABA-2-oxoglutarate aminotransferase. Typical conditions used to localize high-affinity GABA receptors are presented in Table 1. A preincubation (to remove endogenous GABA) for 20 min in 0.31M Tris-citrate, pH 7.1, at ice-bath temperature preceded the actual binding in the same buffer. These conditions were chosen only after Palacios et al. (1980) had determined that they were labeling the authentic GABA receptor through preliminary kinetic (K_d = 6.5 nM) and pharmacological analyses. Under the conditions used for autoradiography, approximately 43% of the high-affinity receptors were labeled.

In the rat brain, Palacios and his colleagues (1980, 1981d) found a heterogenous distribution of receptors. White matter

TABLE 2
Autoradiographic Studies of Amino Acid Receptors

Amino acid receptor	Reference
GABA	Palacios et al., 1980
	Palacios et al., 1981b, c, d
	Penney et al., 1981
	Unnerstall et al., 1981
	Wilkin et al., 1981
	Yazulla et al., 1981
	Palacios and Kuhar, 1982
	Penney and Young, 1982
	Pan et al., 1983
Glycine	Zarbin et al., 1981
	Whitehouse et al., 1983
Glutamate	Foster et al., 1981
	Henke et al., 1981
	Monaghan and Cotman, 1982
	Berger and Ben-Ari, 1983
	Greenamyre et al., 1983, 1984
	Halpain et al., 1983
	Monaghan et al., 1983a
	Unnerstall and Wamsley, 1983

areas had no detectable receptors. In the telencephalon, high concentrations of high-affinity GABA receptors were seen in the external plexiform layer of the olfactory bulb, whereas moderate levels were present in superficial layers (I–IV) of the neocortex, the molecular layer of the dentate gyrus, and the olfactory tubercle. The caudate-putamen and globus pallidus had low levels of receptors despite high levels of GABA and glutamic acid decarboxylase (GAD), GABA's synthesizing enzyme.

In the diencephalon, high concentrations were present in several thalamic nuclei, including portions of the ventral and geniculate nuclei. The supraoptic nucleus was the only hypothalamic nucleus with a high concentration of high-affinity GABA receptors. The hypothalamus generally had low quantities of receptors. Low levels were present in the retinal inner plexiform layer (Wamsley, Palacios, and Kuhar, personal communication).

No structure in the mesencephalon contained high levels of receptors. The superficial layer of the superior colliculus, cortical layer of the inferior colliculus, nucleus pretectalis, nucleus interstitialis of Cajal, and substantia nigra pars reticulata had intermediate levels. Other brainstem structures contained only low levels except for the granule cell layer of the cerebellum. This structure had the highest level of receptors in the brain with clusters of receptors apparently over the glomeruli.

In general, Palacios and coworkers (1981d) found a fair correlation between concentrations of receptors and GABA and GAD. Some areas, such as the basal ganglia, have high GABA and GAD levels, whereas the receptor was found in low concentrations. The converse was true in other areas, such as some thalamic nuclei. These discrepancies may be explained in terms of spare receptors, different postsynaptic geometries or efficacies, or nonvisualized receptors of lower affinity or different specificity. Wilkin et al. (1981) demonstrated this latter possibility in the cerebellum. They showed autoradiographically previously undetected calcium-dependent, bicuculline-insensitive GABA$_B$ sites in the molecular layer of the cerebellum. Obviously, this will be interesting to pursue in other areas in which GABA levels are high and receptors are relatively low.

Another radiohistochemical study investigated the question of multiple GABA receptors (Palacios et al, 1981b; Unnerstall et al., 1981). They noted that benzodiazepine receptor binding was stimulated by GABA at a 1000-fold greater concentration of GABA than was necessary to label high-affinity GABA receptors (reviewed in Unnerstall et al., 1981). Furthermore, the distributions of high-affinity benzodiazepine and GABA receptors were strikingly different (Young and Kuhar, 1979b, 1980; Palacios et al., 1980, 1981d). Unnerstall and coworkers compared the distributions of high-affinity GABA and benzodiazepine receptors and GABA stimulation of benzodiazepine binding (Fig. 2). They demonstrated that the magnitude of GABA stimulation of benzodiazepine binding did not correlate with the quantity of high-affinity GABA receptors. Instead, there was a positive linear correlation with the quantity of baseline benzodiazepine receptors. This suggested that benzodiazepine receptors are associated with a subclass of GABA binding sites, probably a different one from that labeled at low nanomolar concentrations of GABA.

Other studies have combined lesioning techniques and GABA receptor autoradiography. For example, Palacios et al. (1981c) injected kainic acid into guinea pig cerebella and noticed

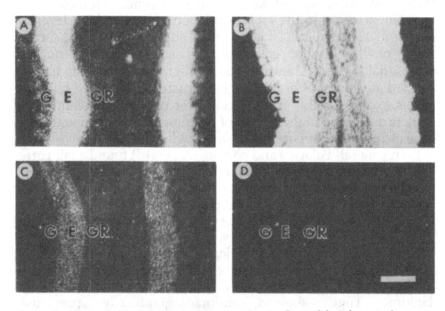

Fig. 2.   Darkfield photomicrographs (reflected brightness is a result of silver grains) of high-affinity benzodiazepine and GABA receptors in adjacent sections of rat olfactory bulb. "A" shows that the binding of ^{3}H-flunitrazepam (1.0 nM in 0.17M Tris-HCl, pH 7.6, at ice-bath temperature for 40 min with 2 min rinse in same buffer) is highest in the external plexiform layer (E) and lower in the glomeruli (G). Very low levels are present in the granule cell layer (GR). A similar distribution is observed for ^{3}H-muscimol (conditions in Table 1) in "C." In "B", 0.2 mM GABA stimulates ^{3}H-flunitrazepam binding in all three layers (49, 67, and 101% increases in G, E, and GR, respectively). "D" is a blank generated with 1.0 nM ^{3}H-flunitrazepam and 1.0 μM clonazepam. Bar = 250 μm (from Palacios et al., 1980; with permission of Raven Press).

loss of benzodiazepine and histamine H_1 receptors by autoradiography, whereas GABA receptors in the granule cell layer were spared. Since only granule cells were spared by kainic acid, this suggested that GABA receptors in the guinea pig cerebellum were largely located on granule cells. An ontogenetic study of the development of GABA receptors in the cerebellum supported this observation as well (Palacios and Kuhar, 1982).

Kainic acid also has been injected into the rat caudate-putamen (Penney et al., 1981; Pan et al., 1983). These authors then measured ^{3}H-muscimol binding autoradiographically and showed no changes in receptor affinity (K_d) in any areas studied. However, they noted significant increases in receptor numbers (B_{max}) in the ipsilateral globus pallidus, entopeduncular nucleus and substantia nigra pars reticulata, suggesting postsynaptic up-regulation. Furthermore, decreases in B_{max} were seen in the ipsilateral anteroventrolateral and ventromedial thalamic nuclei. They concluded that this downstream effect occurred because decreased inhibitory input to the entopeduncular nucleus, substantia nigra, and pars reticulata, allowed activation of a GABAergic projection to the thalamus. Similar findings were observed in a case of Huntington's disease studied by the in vitro autoradiographic technique (Penny and Young, 1982). These studies emphasize the important role that radiohistochemical techniques can play in studying anatomy and receptor physiology, even in the human.

3.2. Glycine Receptors

Glycine is presumed to be the other major inhibitory neurotransmitter in the CNS, especially in the brainstem and spinal cord. Strychnine is a potent antagonist that has been used to define glycine receptors in the brain in vitro (Young and Snyder, 1973; Ferkany, this volume). Zarbin et al. (1981) studied the distribution of glycine receptors using [^{3}H]-strychnine under conditions shown in Table 2. They previously defined the binding site in their tissue sections as the glycine receptor by studying K_d ($k_{-1}/k_1 = 13$ nM; by saturation data, 15 nM), Hill slope (1.1), and pharmacology. Under their conditions of binding for autoradiography, approximately 30% of the high affinity sites were labeled at the end of the incubation.

No telencephalic structures were labeled. The retinal inner plexiform layer had moderate levels (Wamsley, Palacios, and Kuhar, personal communication). In the dicephalon, low levels were found only in some thalamic nuclei, including the parafascicular nuclei and zona incerta. In the mesencephalon, a moderate amount of strychnine binding was observed in the rostral cuneiform nucleus. Low levels were found in other mesencephalic structures, such as the substantia nigra, pars reticulata, superior colliculus, and periaqueductal gray.

No binding was observed in the cerebellum or its nuclei although many pontine nuclei contained moderate to high levels. Structures with high levels included the facial and trigeminal nuclei. Moderate levels were found in the medial vestibular, lateral lemniscal, superior olivary, dorsal cochlear, rostral pontine reticular (pars oralis), and pontine reticular tegmental nuclei. Lower levels were found in many other pontine nuclei.

In the medulla, high strychnine binding was observed in the hypoglossal nucleus, dorsal motor nucleus of the vagus, substantia gelatinosa and spinal tract of the trigeminal nucleus, and lateral reticular nucleus. Moderate levels were found in many other nuclei, including cuneate, gracile, ambiguous, and various reticular nuclei.

The spinal cord displayed the highest receptor levels in the CNS. The receptor levels were especially prominent in laminae II and III of the dorsal horn and lamina VII of the ventral horn at all levels. Lamina V also had high levels, especially in the cervical cord. Interestingly, all laminae except I had high levels of strychnine binding in the cervical cord, whereas levels in the thoracic cord were more moderate (except in laminae II, III, and VII). In the lumbar enlargement, high densities again were found in the ventral horn.

The autoradiographic distribution of glycine receptors in human CNS has not been studied extensively. However, strychnine binding has been studied radiohistochemically in the spinal cord in postmortem tissue from persons with amyotrophic lateral sclerosis (a degenerative disease with loss of upper and lower motor neurons) and controls (Whitehouse et al., 1983). In the cervical cord of the control group (three with no neurological disease, two with Huntington's disease, and one with multiple sclerosis), the highest density of glycine receptors was found in laminae II and III with a slightly lower level in lamina IX. A similar picture was presented for the lumbar cord (Fig. 3). In the patients with amyotrophic lateral sclerosis, a significant decrease (31%) in glycine receptors occurred only in the ventral horn, the area which experiences a reduction (75–80%) in neurons.

The glycine receptor distribution correlates well with that of glycine itself and its high-affinity uptake sites, in contrast with the fair correlation of distribution of GABA receptors with GABA, GAD, or GABA uptake (discussed above).

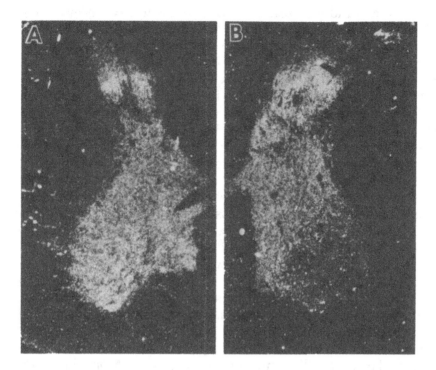

Fig. 3. Autoradiographic localization of glycine receptors in the lumbar spinal cord of a control patient (A) and amyotrophic lateral sclerosis (ALS) patient (B). The highest receptor (grain) densities are found in the dorsal horn, laminae II and III. Compared with controls, the major reduction in receptors is found in the ventral horn region containing motor neurons (from Whitehouse, et al., 1983; with permission).

3.3. Glutamic Acid Receptors

Glutamic acid is considered a principal excitatory amino acid in the CNS, and high-affinity binding sites have been described in membrane homogenate preparations. In addition, several classes of glutamate receptors have been proposed on the basis of various compounds' abilities to displace [^{3}H]-L-glut amate and on the basis of electrophysiological studies. N-Methyl-D-aspartate (NMDA), kainic acid (KA), quisqualic acid, and 2-amino-4-phosphonobutyric acid (APB) have been used to define four receptors (for references and discussion, see J.W. Ferkany, this volume, and Table 2).

Autoradiographic studies of these receptors are just beginning. The KA receptor initially was studied autoradiographically (Foster et al., 1981; Henke et al., 1981; Monaghan and Cotman,

1982; Berger and Ben-Ari, 1983; Unnerstall and Wamsley, 1983). Monaghan and Cotman (1982) mapped the distribution of high- and low-affinity KA receptors in the rat CNS after preliminary verification of the presence of a site with a K_d of 69 nM with appropriate pharmacological specificity. Assuming equilibrium conditions and that the high affinity site had a K_d of approximately 10 nM, these authors labeled approximately 90% of the high- and 60% of the low-affinity sites with 100 nM ^{3}H-KA before the 1 min wash. Unnerstall and Wamsley (1983) performed similar preliminary biochemical studies and chose 15 nM ^{3}H-KA, which labeled 55% of the high-affinity and 19% of the low-affinity sites in their preparations after washing. Despite the different conditions used, both groups described essentially identical distributions with a few exceptions that may be explained by the additional labeling of low-affinity sites by Monaghan and Cotman.

In the telencephalon, high levels were observed over the external plexiform and granule cell layers of the olfactory bulb. The neocortex had elevated grain densities over laminae I, V, and VI, and this trilaminar pattern especially was well defined in frontal and cingulate regions. Unnerstall and Wamsley found moderate grain densities in the superficial layers of the pyriform cortex and deeper layers of the entorhinal cortex, but Monaghan and Cotman saw a more uniform distribution in these areas. The olfactory tubercle had elevated levels in the polymorphic layer. Moderate levels were seen in the amygdala. In the basal ganglia, high uniform labeling was seen over the striatum and nucleus accumbens septi. In contrast, no binding was seen in the globus pallidus.

The septal and septofimbrial nuclei, parts of the lateral septal nucleus, and bed nucleus of the stria terminalis had moderate amounts of KA receptors. These studies confirmed the findings of Foster et al. (1981) who found very high levels in the stratum lucidem of CA3 and CA4 in the hippocampal formation. Similar results were reported by Berger and Ben-Ari (1983) and in a complementary fashion by Monaghan et al. (1983a) using ^{3}H-L-glutamate (see below). These regions receive the termination of granule cell mossy fibers, but the exact location of the KA receptor in relation to the synapse is unclear. Lesions of either the granule cells (presynaptic component) or of the pyramidal cells (postsynaptic component) resulted in loss of stratum lucidem KA receptors (Monaghan and Cotman, 1982).

In the diencephalon, moderate levels were observed in the thalamic reticular nucleus, zona incerta, and hypothalamus, except for the median eminence, which had a high level of receptors. The posterior pituitary also contained high levels. Only low to moderate levels were found in the rest of the brain, although Monaghan and Cotman saw a high density in the cerebellar granule cell layer. Henke et al. (1981) made a similar observation in the pigeon cerebellum. However, they also observed an abundant lower-affinity site in the molecular layer that exhibited positive cooperativity and which, apparently, does not exist in mammals.

Initially, investigators assumed that ^{3}H-L-glutamate would not be an appropriate ligand because of to the glutamate receptor's low affinity for this ligand: approximately 1 μM in some laboratories (however, see Biziere et al., 1980). However, several groups recently have confirmed biochemically that the glutamate receptor survives in sections and have produced autoradiograms using ^{3}H-L-glutamate (Greenamyre et al., 1983; Halpain et al., 1983; Monaghan et al., 1983a). Halpain and coworkers found a K_d of 500 nM and labeled approximately 23% of the binding sites before washing. However, since they used a 10 min rinse, negligible binding should have remained (assuming a $t_{1/2}$ for dissociation of about 30 s). Consequently, their autoradiograms are difficult to interpret.

Greenamyre et al. (1983) found a K_d of about 1 μM and labeled approximately 9% of the receptors at equilibrium and lost about 25% during their 10 s wash. Under these conditions, they found high densities in the cortex, striatum, stratum moleculare of the hippocampus, and molecular layer of the cerebellum—each an area thought to receive extensive afferents using glutamate. Portions of the thalamus also appeared to contain moderate to high levels. A more detailed presentation by these authors (Greenamyre et al., 1984) appeared after this manuscript went to press.

Monaghan and coworkers (1983a) employed conditions similar to those of Greenamyre et al. (1983) to study different conditions in consecutive sections through the hippocampus. They demonstrated four classes of binding sites. In the absence of calcium, chloride, and sodium ions, they observed a large population of NMDA-sensitive receptors, predominantly in the stratum oriens and stratum radiatum of the CA1 field and inner portions of the dentate gyrus molecular layer. The Schaffer collaterals project to the CA1 sites and the CA4 system to the molecular layer of dentate gyrus, and all are thought to use glutamate. Moderate levels were located elsewhere in the hippocampus. NMDA-insensitive, KA-displaceable ^{3}H-glutamate sites were found in the stratum lucidum of CA3 and commissural/associational layer of the dentate gyrus, in agreement with ^{3}H-KA autoradiography (see above). A third site found in the absence of calcium, chloride, and sodium ions was located in the pyramidal cell layer of CA1 and CA3 and was sensitive to quisqualate, α-amino-3-hydroxy-5-methylisoxazole-4-pr opionic acid (AMPA), and L-serine-o-sulfate. The fourth site was evident in the presence of calcium and chloride, was predominant in the stratum lacunosum-moleculare and in the dentate gyrus molecular layer and presumably represents the APB site (Monaghan et al., 1983b). At the present time, the significance of these four sites is not completely understood and awaits further electrophysiological and biochemical study. Although distributions of glutamic acid uptake sites, glutaminase, and glutamic acid tend to parallel most receptor populations, the actual transmitter active at these sites is unknown. Glutamic acid may be a neurotransmitter, but it is certainly possible that various short peptides containing glutamic acid or aspartic acid may, in fact, be the predominant neurotransmitters (Zaczek et al., 1983).

4. Amino Acid Uptake Sites

The in vitro autoradiographic approach to localization of amino acid uptake sites has received less attention than the in vivo one. The former approach, however, has several potential advantages. These include control of uptake conditions and metabolism, different conditions applied to adjacent sections or slices, and use of human postmortem tissue. This approach allows biochemical definition of the uptake sites in a fashion similar to the one used to study amino acid receptors radiohistochemically. This is especially important since the amino acids are potential neuronal metabolic substrates and may be transported into compartments other than those for subsequent neurotransmitter release. Also, uptake may occur into nonneuronal cells and one must avoid measuring receptor sites as well. In addition to these biochemical studies, the processing for autoradiography should be assessed to assure that no loss of ligand by diffusion occurs, just as for receptor autoradiography.

Although the studies listed in Table 3 generally report an association of radioactivity with neuronal elements, none has demonstrated that receptors are not also labeled, or that diffusion from the uptake site is absent during the fixation and processing for autoradiography. This is an especially troubling concern for electron microscopic evaluation. Similarly, the majority of reports have not defined the nature of the accumulated radioactivity in terms of kinetics, pharmacology, or even chemical form.

TABLE 3
In Vitro Autoradiographic Studies of Amino Acid Uptake
Sites in Tissue Sections and Slices

Amino acid	Reference
GABA	Hökfelt and Ljungdahl, 1970
	Iversen and Bloom, 1972
	Hattori et al., 1973
	Marshall and Voaden, 1975
	Hampton and Redburn, 1983
	Sarthy, 1983
Glycine	Hökfelt and Ljungdahl, 1971
	Matus and Dennison, 1971
	Iversen and Bloom, 1972
	Wilkin et al., 1981
	Sarthy, 1983
Taurine	Schulze and Neuhoff, 1983
Aspartate	Ehinger, 1981
	Parsons and Rainbow, 1983
Glutamate	DeBarry et al., 1982
	Hampton and Redburn, 1983

The purpose of this discussion is not to criticize the results obtained in the reports listed in Table 3, for they may be entirely valid, but to emphasize the necessary controls and opportunities for further study. In this light, the paper by Parsons and Rainbow (1983) presents a potentially important alternative approach to the study of uptake sites by autoradiography. They proposed that the sodium-dependent ^{3}H-D-aspartate binding sites they studied were, in fact, uptake sites and presented biochemical evidence to support this notion. In a similar fashion, it may be possible to label and use various specific amino acid uptake inhibitors and avoid some of the difficulties encountered when one uses the amino acids themselves. This approach has been used by the same group to study norepinephrine and serotonin uptake sites (Rainbow et al., 1982a; Biegon and Rainbow, 1983). Obviously, additional questions, such as the relevance of these binding sites in once-frozen tissue sections, need further investigation.

5. Summary

I have presented an approach to autoradiographic localization of receptors at the light microscopic level. The important basic aspects of the approach are the preliminary biochemical characterization and subsequent avoidance of diffusion of the ligand for the autoradiography. And, although I have presented a survey mainly of amino acid receptor autoradiography, the principles employed can be applied just as well to uptake sites. These radiohistochemical techniques should continue to further our understanding of neuronal systems in not only experimental animals, but also in humans. The application of these methods to the study of human disease (Penny and Young, 1982; Whitehouse et al., 1983) is just beginning to bear fruit and certainly promises to be an exciting future direction of research.

Acknowledgments

I would like to express my gratitude to Dr. Frederick Wooten for his critical review of this manuscript, to Dr. George Alheid for his suggestion concerning ethidium bromide, and to Ms. Rose Powell and Mrs. Martha Smith for secretarial assistance. Development of some of the autoradiographic procedures described above was supported in part by US Public Health Service Grant MH07624.

References

Alexander G. M., Schwartzman R. J., Bell R. D., Yu J., and Renthal A. (1981) Quantitative measurement of local cerebral metabolic rate for glucose utilizing tritiated 2-deoxyglucose. *Brain Res.* **223**, 59–67.

Berger M. and Ben-Ari Y. (1983) Autoradiographic visualization of [^{3}H] kainic acid receptor subtypes in the rat hippocampus. *Neurosci. Lett.* **39**, 237–242.

Biegon A. and Rainbow T. C. (1983) Localization and characterization of [^{3}H] desmethylimipramine binding sites in rat brain by quantitative autoradiography. *J. Neurosci.* **5**, 1069–1076.

Biziere K., Thompson H., and Coyle J. T. (1980) Characterization of specific, high-affinity binding sites for L-[^{3}H]-glutamic acid in rat brain membranes. *Brain Res.* **183**, 421–433.

Cuatrecasas P. and Hollenberg M. D. (1976) Membrane receptors and hormone action. *Adv. Protein Chem.* **30**, 251–451.

DeBarry J., Langley O. K., Vincendon G., and Gombos G. (1982) L-Glutamate and L-glutamine uptake in adult rat cerebellum: an autoradiographic study. *Neuroscience* **7**, 1289–1297.

Ehinger B. (1981) [^{3}H]-D-Aspartate accumulation in the retina of pigeon, guinea pig and rabbit. *Exp. Eye Res.* **33**, 381–391.

Foster A. C., Mena E. E., Monaghan D. T., and Cotman C. W. (1981) Synaptic localization of kainic acid binding sites. *Nature* **289**, 73–75.

Geary W. A., II, and Wooten G. F. (1983) Quantitative film autoradiography of opiate agonist and antagonist binding in rat brain. *J. Pharmacol. Exp. Ther.* **225**, 234–240.

Greenamyre J. T., Young A. B., and Penney J. B. (1984) Quantitative autoradiographic distribution of L-[^{3}H]-glutamate-binding sites in rat central nervous system. *J. Neurosci.* **4**, 2133–2144.

Greenamyre J. T., Young A. B., and Penney J. B. (1983) Quantitative autoradiography of L-[^{3}H] glutamate binding to rat brain. *Neurosci. Lett.* **37**, 155–160.

Halpain S., Parsons B., and Rainbow T.C. (1983) Tritium-film autoradiography of sodium-independent glutamate binding sites in rat brain. *Eur. J. Pharmacol.* **86**, 313–314.

Hampton C. K. and Redburn D. A. (1983) Autoradiographic analysis of ^{3}H-glutamate, ^{3}H-dopamine and ^{3}H-GABA accumulation in rabbit retina after kainic acid treatment. *J. Neurosci. Res.* **9**, 239–251.

Hattori T., McGeer P. L., Fibiger H. C., and McGeer E. G. (1973) On the source of GABA-containing terminals in the substantia nigra. Electron microscopic autoradiographic and biochemical studies. *Brain Res.* **54**, 103–114.

Henke H., Beaudet A., and Cuénod M. (1981) Autoradiographic localization of specific kainic acid binding sites in pigeon and rat cerebellum. *Brain Res.* **219**, 95–105.

Herkenham M., and Pert C. B. (1982) Light microscopic localization of brain opiate receptors: a general autoradiographic method which preserves tissue quality. *J. Neurosci.* **2**, 1129–1149.

Hökfelt T. and Ljungdahl Å. (1970) Cellular localization of labeled gamma-aminobutyric acid (^{3}H-GABA) in rat cerebellar cortex: an autoradiographic study. *Brain Res.* **22**, 391–396.

Hökfelt T. and Ljungdahl Å. (1971) Light and electron microscopic autoradiography on spinal cord slices after incubation with labeled glycine. *Brain Res.* **32**, 189–194.

Iversen L. L. and Bloom F. E. (1972) Studies of the uptake of ^{3}H-GABA and ^{3}H-glycine in slices and homogenates of rat brain and spinal cord by electron microscopic autoradiography. *Brain Res.* **41**, 131–143.

Kuhar M. J. (1982) Localization of drug and neurotransmitter receptors in brain by light microscopic autoradiography, in *Handbook of Psychopharmacology* (Iversen L.L., Iversen S.D., and Snyder S.H., eds.), pp. 299–320. Plenum Press, New York.

Kuhar M. J. (1983) Autoradiographic localization of drug and neurotransmitter receptors, in *Handbook of Chemical Neuroanatomy. Vol. 1:* Methods of Chemical Neuroanatomy. (Bjorklund A. and Hökfelt T., eds.), pp. 398–415. Elsevier, New York.

Kuhar M. J. (1984) Receptor localization with the microscope, in *Neurotransmitter Receptor Binding* (Yamamura H.I., Enna S.J., and Kuhar M.J., eds.), in press, Raven Press, New York.

Kuhar M. J. and Unnerstall J. R. (1982) In vitro labeling receptor autoradiography: loss of label during ethanol dehydration and preparative procedures. *Brain Res.* **244**, 178–181.

Lysz T., Toga A. W., and Geary W. A., II. (1982) Standardization of ³H-tissue images with different isotopic standards. *Soc. Neurosci. Abstract* **8,** 645.

Marshall J. and Voaden M. (1975) Autoradiographic identification of the cells accumulating ³H-γ-aminobutyric acid in mammalian retinae: a species comparison. *Vision Res.* **15,** 459–461.

Matus A. I. and Dennison M. E. (1971) Autoradiographic localization of tritiated glycine at "flat-vesicle" synapses in spinal cord. *Brain Res.* **32,** 195–197.

Möhler H., Richards J. G., and Wu J.-Y. (1981) Autoradiographic localization of benzodiazepine receptors in immunocytochemically identified γ-aminobutyric synapses. *Proc. Natl. Acad. Sci. USA* **78,** 1935–1938.

Monaghan D. T. and Cotman C. W. (1982) The distribution of [³H] kainic acid binding sites in rat CNS as determined by autoradiography. *Brain Res.* **252,** 91–100.

Monaghan D. T., Holets V. R., Toy D. W., and Cotman C. W. (1983a) Anatomical distributions of four pharmacologically distinct ³H-L-glutamate binding sites. *Nature* **306,** 176–179.

Monaghan D. T., McMills M. C., Chamberlin A. R., and Cotman C. W. (1983b) Synthesis of [³H]2-amino-4-phosphonobutyric acid and characterization of its binding to rat brain membranes: a selective ligand for the chloride/calcium-dependent class of L-glutamate binding sites. *Brain Res.* **278,** 137–144.

Palacios J. M. and Kuhar M. J. (1982) Ontology of high-affinity GABA and benzodiazepine receptors in rat cerebellum: an autoradiographic study. *Brain Res.* **2,** 531–539.

Palacios J. M., Young W. S., III, and Kuhar M. J. (1980) Autoradiographic localization of GABA receptors in the rat cerebellum. *Proc. Natl. Acad. Sci. USA* **77,** 670–674.

Palacios J. M., Niehoff D. L., and Kuhar M. J. (1981a) Receptor autoradiography with tritium-sensitive film: potential for computerized densitometry. *Neurosci. Lett.* **25,** 101–105.

Palacios J. M., Unnerstall J. R., Young W. S., III, and Kuhar M. J. (1981b) Radiohistochemical studies of benzodiazepine and GABA receptors and their interactions, in *GABA and Benzodiazepine Receptors* (Costa E., Di Chiara G., and Gessa G.L., eds.), pp. 53–60. Raven Press, New York.

Palacios J. M., Wamsley J. K. and Kuhar M. J. (1981c) GABA, benzodiazepine, and histamine-H₁ receptors in the guinea pig cerebellum: effects of kainic acid injections studied by autoradiographic methods. *Brain Res.* **214,** 155–162.

Palacios J. M., Wamsley J. K., and Kuhar M. J. (1981d) High affinity GABA receptors-autoradiographic localization. *Brain Res.* **222,** 285–307.

Pan H. S., Frey K. A., Young A. B., and Penney J. B., Jr. (1983) Changes in [³H] muscimol binding in substantia nigra, entopeduncular nucleus, globus pallidus, and thalamus after striatal lesions as demonstrated by quantitative receptor autoradiography. *J. Neurosci.* **3**, 1189–1198.

Parsons B. and Rainbow T. C. (1983) Quantitative autoradiography of sodium-dependent [^{3}H]D-aspartate binding sites in rat brain. *Neurosci. Lett.* **36**, 9–12.

Penney J. B., Jr., Pan H. S., Young A. B., Frey K. A., and Dauth G. W. (1981) Quantitative autoradiography of [^{3}H] muscimol binding in rat brain. *Science* **214**, 1036–1038.

Penney J. B. and Young A. B. (1982) Quantitative autoradiography of neurotransmitter receptors in Huntington disease. *Neurology* **32**, 1391–1395.

Rainbow T. C., Biegon A., and McEwen B. S. (1982a) Autoradiographic localization of imipramine binding in rat brain. *Eur. J. Pharmacol.* **77**, 363–364.

Rainbow T. C., Bleisch W. V., Biegon A., and McEwen B. S. (1982b) Quantitative densitometry of neurotransmitter receptors. *J. Neurosci. Meth.* **5**, 127–138.

Rogers A. W. (1979) *Techniques of Autoradiography.* Elsevier, New York.

Roth L. J., Diab I. M., Watanabe M., and Dinerstein R. (1974) A correlative radioautographic fluorescent and histochemical technique for cytopharmacology. *Mol. Pharmacol.* **10**, 986–998.

Sarthy P. V. (1983) Release of [^{3}H]γ-aminobutyric acid from glial (Muller) cells of the rat retina: effects of K$^+$, veratridine, and ethylenediamine. *J. Neurosci.* **12**, 2494–2503.

Schulze E. and Neuhoff V. (1983) Uptake, autoradiography and release of taurine and homotaurine from retinal tissue. *Int. J. Neurosci.* **18**, 253–268.

Taylor, N. R., Unnerstall, J. R., Mashal, R. D., DeSouza, E. B., and Kuhar, M. J. (1984) Receptor autoradiography: coping with regional differences in autoradiographic efficiency with tritium. *Soc. Neurosci. Abs.* **10**, 557.

Unnerstall J. R. and Wamsley J. K. (1983) Autoradiographic localization of high-affinity [^{3}H]-kainic acid binding sites in the rat forebrain. *Eur. J. Pharmacol.* **86**, 361–371.

Unnerstall J. R., Kuhar M. J., Niehoff D. L. and Palacios J. M. (1981) Benzodiazepine receptors are coupled to a subpopulation of γ-aminobutyric acid (GABA) receptors: evidence from a quantitative autoradiographic study. *J. Pharmacol. Exp. Ther.* **218**, 797–804.

Unnerstall J. R., Niehoff D. L., Kuhar M. J. and Palacios J. M. (1982) Quantitative receptor autoradiography using [^{3}H]Ultrofilm: application to multiple benzodiazepine receptors. *J. Neurosci. Meth.* **6**, 59–73.

Wamsley J. K. and Palacios J. M. (1983) Autoradiographic distribution of receptor ligand binding–amino acid and benzodiazepine receptors. *Handbook of Chemical Neuroanatomy Vol. 2–Classical Transmitters in the CNS.* (Bjorklund A., Hökfelt T., and Kuhar M.J., eds.), in press, Elsevier, New York.

Whitehouse P. J., Wamsley J. K., Zarbin M. A., Price D. L., Tourtellotte

W.W., and Kuhar M.J. (1983) Amyotrophic lateral sclerosis: alterations in neurotransmitter receptors. *Ann. Neurol.* **14**, 8–16.

Wilkin G. P., Csillag A., Balázs R., Kingsbury A. E., Wilson J. E., and Johnson A. L. (1981) Localization of high affinity [^{3}H] glycine transport sites in the cerebellar cortex. *Brain Res.* **216**, 11–33.

Wilkin G. P., Hudson A. L., Hill D. R., and Bowery N. G. (1981) Autoradiographic localization of GABA$_B$ receptors in rat cerebellum. *Nature* **294**, 584–587.

Yamamura H. I., Enna S. J. and Kuhar M. J., eds. (1984) *Neurotransmitter Receptor Binding.* In press, Raven Press, New York.

Yazulla S. (1981) GABAergic synapses in the goldfish retina: an autoradiographic study of ^{3}H-musciomol and ^{3}H-GABA binding. *J. Comp. Neurol.* **200**, 83–93.

Young A. B. and Snyder S. H. (1973) Strychnine binding associated with glycine receptors of the central nervous system. *Proc. Natl. Acad. Sci. USA* **70**, 2832–2836.

Young W. S., III and Kuhar M. J. (1979a) A new method for receptor autoradiography: [^{3}H] opioid receptors in rat brain. *Brain Res.* **179**, 255–270.

Young W. S., III and Kuhar M. J. (1979b) Autoradiographic localization of benzodiazepine receptors in the brains of humans and animals. *Nature* **280**, 393–395.

Young W. S., III and Kuhar M. J. (1980) Radiohistochemical localization of benzodiazepine receptors in rat brain. *J. Pharmacol. Exp. Ther.* **212**, 337–346.

Young W. S., III and Kuhar M. J. (1981) The light microscopic radiohistochemistry of drugs and neurotransmitter receptors using diffusible ligands, in *Current Trends in Morphological Techniques, Vol. III.* (Johnson J.E., Jr., Adelman R.C. and Roth G.S., eds.), pp. 119–135. CRC Press. Boca Raton, FL.

Zaczek R., Koller K., Cotter R., Heller D., and Coyle J.T. (1983) *N*-Acetylaspartylglutamate: an endogenous peptide with high affinity for a brain. *Proc. Natl. Acad. Sci. USA* **80**, 1116–1119.

Zarbin M. A., Wamsley J. K., and Kuhar M.J. (1981) Glycine receptor: light microscopic autoradiographic localization with [^{3}H] strychnine. *J. Neurosci.* **1**, 532–547.

Chapter 9

Determination of Transmitter Amino Acid Turnover

Frode Fonnum

1. Introduction

Five amino acids have received considerable attention as putative neurotransmitters in mammalian brain; namely, γ-aminobutyric acid (GABA), glycine, glutamate, aspartate, and taurine (Fonnum, 1978).

A large body of evidence supports the concept that GABA is an important inhibitory neurotransmitter (Okada and Roberts, 1982). Glycine is a strong transmitter candidate at the strychnine-sensitive inhibitory synapses in the medulla and the spinal cord (Aprison and Nadi, 1978). Glutamate is ubiquiously distributed in the brain and occupies a central position in brain metabolism. It was shown early to have excitatory action on neurons, and evidence collected during the last eight years indicates that it may be quantitatively the most important transmitter of cortical efferent fibers (Fonnum, 1984).

Aspartate is closely related metabolically and chemically to glutamate. In many respects it is difficult to differentiate between aspartate and glutamate as neurotransmitter candidates (Fonnum, 1984). There are, however, some fibers and terminals in which aspartate is regarded as a stronger transmitter candidate than glutamate. These include the cerebellar climbing fibers (Wiklund et al., 1982), hippocampal commissural fibers (Nadler et al., 1978), olfactory tract (Collins and Probett, 1981), cochlear nucleus afferents (Wenthold, 1979), and spinal cord interneurons (Davidoff et al., 1967).

Although the taurine level in brain is almost as high as that of glutamate, its role as a neurotransmitter is being questioned

201

(Rassin et al., 1981). Taurine is released from brain in a manner more consistent with that of a membrane modulator than a neurotransmitter (Guiener et al., 1975; Collins, 1977). Since there are no data on its turnover in brain tissue, there will be none included in this review.

2. The Level of Transmitter Amino Acids in the Brain

There are large variations in the regional distribution pattern of the four transmitter amino acids (Costa et al., 1979; van der Heyden et al., 1979; Fahn, 1976; Johnson, 1978). One would expect the level of these amino acids to reflect their quantitative importance as neurotransmitters, i.e., the higher the level, the higher the number of terminals that release it. The level of aspartate, glutamate, and glycine does not change markedly postmortem, whereas GABA increases very rapidly after death. This means that special precaution must be taken before the GABA level can be assayed, whereas little precaution needs to be considered before assaying the three other amino acids. GABA can be determined after microwave irridation (Knierem et al., 1977) or after giving a glutamate decarboxylase (GAD)-inhibitor such as 3-mercaptopropionic acid (Karlsson et al., 1974; van der Heyden and Korf, 1978).

The transmitter amino acids are present in the brain in millimolar concentrations, whereas the transmitter amines, including acetylcholine, are present in micromolar concentrations (Smith et al., 1982). Although the transmitter amines are specifically localized in the synaptosome and vesicle fractions, the transmitter amino acids are mainly recovered in the nonsynaptosomal cytoplasm (Mangan and Whittaker 1966; Rassin, 1972; Kontro et al., 1980). Whether this result is an artifact has been frequently discussed. Recent studies on the effects of different drugs on the distribution of GABA between the soluble and synaptosome fractions have indicated that specific differences between the two fractions can be seen (Wood et al., 1980; Wood, 1981; Wood and Kurylo, 1984). Lesions of specific aspartergic, glutamergic fibers lead to a specific loss of that amino acid in the terminal region of the brain (Collins and Probett, 1981; Fonnum et al., 1981; Aprison and Nadi, 1978). Also, subcellular fractionation after lesion of aspartergic or glutamergic fibers shows specific loss of the neurotransmitter amino acid in the synaptosome fraction (Nadler and Smith, 1981; Fonnum et al., unpublished). These findings indicate that transmitter amino acids are highly localized in specific nerve terminals. This does not necessarily indicate that glutamate, aspartate, and glycine are not present in all other terminals, however—they may occur at lower concentrations. The local concentration of GABA in GABA-ergic terminals has been estimated by several different methods to be very high, possibly between 50–150 mM (Fonnum and Walberg, 1973; Okada, 1982). There is extensive regional variation in the GABA levels, and good correlation between the GABA level and GAD activities (Baxter, 1976). The glycine levels are less variable, but that amino acid is present at high levels in the regions where it is believed to have an important transmitter function (Aprison and Nadi, 1978). Glutamate, on the other hand, is present in a high concentration all over the brain, and the difference between the highest and lowest level is only twofold (Lane et al., 1982). The regional variation of aspartate levels is higher than that of glutamate levels (Davidoff et al., 1967). The changes in the level of a neurotransmitter are often poor indicators of the turnover rate. A decrease may, however, indicate that the tissue cannot maintain the transmitter synthesis rate.

The turnover of a neurotransmitter can be determined when its synthesis or metabolism is well defined (Fonnum, 1981). For the amino acid transmitters, this seems to be the case with GABA only. In the following, therefore, the methods for determining the turnover of GABA will be discussed in detail. Glutamate has a double role in amino acid neurotransmission, both as a precursor for GABA and as a neurotransmitter. There are, however, few data on the turnover of transmitter glutamate; therefore a brief discussion of the possibilities that exist will be given. Aspartate is closely linked to glutamate and is the most difficult amino acid to study kinetically. At present, nobody has been able to fit the behavior of aspartate in brain into a kinetic model (Clarke et al., 1978). This is in part related to the very low level of oxaloacetic acid (van den Berg and Garfinkel, 1971), since the level of aspartate in the brain is probably regulated through an equilibrium between aspartate, glutamate, oxaloacetate, and 2-oxoglutarate. Thus the level of aspartate may change as a result of the metabolic status of the animal. Hypoglycemia favors the formation of aspartate in part at the expense of glutamate (Butterworth et al., 1982). The few data that exist on the turnover of glycine will also be reviewed.

3. Compartmentation of Glutamate-Glutamine-GABA in the Brain and Localization of the Corresponding Enzymes

Kinetic studies on the metabolism of amino acid precursors in the brain have shown that the synthesis and metabolism of glutamate, aspartate, glutamine, and GABA can be separated into a two-compartment model (Berl et al. 1961, Clarke et al., 1978; Berl and Clarke, 1978; van den Berg and Garfinkel, 1971; Berl et al., 1970; Machiyama et al., 1970; Balazs et al., 1970). The two compartments are called the large and small glutamate compartment, respectively, and each contains its own tricarboxylic acid (TCA) cycle. The compartmentation is easily demonstrated by comparing the specific radioactivity of glutamine to that of glutamate after injection of labeled precursors. In the case of $[^{14}C]$-bicarbonate, short chain fatty acids (acetate, propionate, butyrate), and amino acids (glutamate, aspartate, GABA, leucine), the ratio exceeds 1.0 a short time after injection. In the case of glucose, pyruvate, lactate, glycerol, and hydroxybutyrate, the ratio stays well below 1.0 (Table 1). The results can be explained if the first group of precursors are assumed to be either preferentially taken up or metabolized in a small pool of glutamate that is responsible for the synthesis of glutamine. The other group of precursors are preferentially taken up or metabolized in a large pool of glutamate, where the synthesis of glutamine is negligible. It is not surprising, considering the heterogeneity of brain tissue, that

TABLE 1

Precursors for the Large and Small Glutamate
Compartments in the Brain

Small compartment	Large compartment
Acetate	Glucose
Propionate	Pyruvate
Butyrate	Lactate
Glutamate	Glycerol
Aspartate	β-Hydroxybutyrate
GABA	
Citrate	
Bicarbonate	
Leucine	
Succinate	

glutamate metabolism can be separated into two pools. The brain contains several different types of neurons and glial cells. The neurons are functionally different with different neurotransmitters and, therefore with different rates of amino acid metabolism. One would thus expect the metabolism of glutamate to be accounted for only by a several-compartment model, and in fact van den Berg et al. (1975) have suggested a five-compartment model. It is therefore likely that each of the two compartments probably results from the summation of a number of different units with a similar metabolic pattern. The compartments are not static and may depend on the experimental conditions. A list of compounds and concentrations used to simulate the two compartment models are given by van den Berg and Garfinkel (1971).

It is not necessary for the two kinetic compartments to correspond to distinct morphological compartments, but that has been shown to be the case. Evidence is accumulating that the small compartment corresponds to glial cells. Autoradiographs of rat dorsal root ganglion show labeling mainly in the satellite glial cells after incubation with [^{14}C]-acetate (Minchin and Beart, 1974). Glutamine synthetase (E.C.6.3.1.2), the enzyme responsible for glutamine synthesis, has been localized in glial cells by an immunohistochemical technique (Martinez-Hernandez et al., 1977). Ultrastructural studies have shown the enzyme to be present in astrocytes (Norenberg and Martinez-Hernandez 1979), and high levels of glutamine synthetase have been found in astrocyte cultures and in isolated astrocyte fractions (Patel, 1982), but not in oligodendrocyte cultures (Schousboe et al., 1977; Raff et al., 1983). This conclusion is also in agreement with the increase of acetate labeling in slices from substantia nigra after lesion which induce gliosis (Minchin and Fonnum, 1979). Degeneration of nerve terminals is often accompanied by astrocyte formation and an increase in glutamine synthetase and glutamine levels (Sandberg et al., 1985).

On the other hand, numerous studies support the concept that the large glutamate compartment is the neuron. Autoradiographs of the dorsal ganglion after incubation with [^{14}C]-glucose show most of the labeling over neuronal cell bodies (Minchin and Beart, 1974). A decrease in amino acids from labeling with glucose has been found in tissues with degenerated neural structures (Minchin and Fonnum, 1979; Nicklas, 1983). The specific localization of GAD to neuronal structures, particularly the GABAergic neuron, is also consistent with the large glutamate compartment being equivalent to neuronal structures (McLaughlin et al., 1974).

The other enzymes involved in glutamate metabolism are usually present in both glial and neuronal structures, often with a specific cellular preference. Aspartate aminotransferase (E.C.2.6.1.1. AspT) is localized in the brain as two isoenzymes with different kinetic properties and with a cytoplasmic and mitochondrial localization, respectively (Fonnum, 1968). After degeneration of certain neuronal tracts, such as the habenula-interpeduncular tract (Sterri and Fonnum, 1980), there is a reduction of AspT activity in the terminal region. This should at least indicate a preferential localization of the enzyme activity in neurons. Immunohistochemical studies of the cytoplasmic isoenzyme have shown that the enzyme is localized to the auditory nerve, photoreceptors, ganglion cells in retina, basket stellate cells in cerebellum, and cells in neocortical layers II and III (Altschuler et al., 1981, 1982; Wenthold and Altschuler, 1983). Therefore, some of these studies suggest that the enzyme may be preferentially localized in aspartate neurons. There is, however, no correlation between the aspartate level and total AspT activity (AspT) in some regions of the spinal cord where the aspartate level varies considerably (Graham and Aprison, 1969). The finding that aminooxyacetic acid in high concentration inhibits glutamate, but not glutamine, formation accords with a localization of the enzyme mainly to the large compartment (Berl and Clarke, 1978). Further, β-methylene-DL_1-aspartate, an inhibitor of AspT, inhibits oxygen consumption from glucose or glutamate to the same extent as AspT inhibition (Fitzpatrick et al., 1983).

The phosphate-stimulated glutaminase (E.C.3.5.1.2.) has been localized to mitochondria, particularly the synaptosome mitochondria, by subcellular fractionation techniques (Salganicoff and De Robertis, 1965; Bradford and Ward, 1976). The results of lesion studies are unequivocal with regard to the specific localization in different types of neurons. Cortical ablation is accompanied by 20% loss of glutaminase activity in neostriatum (Ward et al., 1982; Walker and Fonnum, unpublished), and intrastriatal injection of kainic acid, by 50% reduction in neostriatum (McGeer and McGeer, 1978; Nicklas et al., 1979). Glutaminase has been shown to be present in higher concentration in enriched Purkinje and granule cell fractions than in astrocyte fractions from cerebellum (Patel, 1982). Immunohistochemical studies showed specific localization of glutaminase activity in certain amacrine cells in retina, mossy fibers and pyramidal cells in hippocampus, granule cells in cerebellum, and cells of neocortical layers V and VI (Wenthold and Altschuler, 1983), all of which probably use glutamate as a transmitter. In conclusion, phosphate-stimulated glutaminase seems to be more dominant in neuronal than glial structures. It is difficult to relate the enzyme to one transmitter type of neuron only, but it is obviously highly localized to at least some GABAergic and glutamergic neurons. It may not be the localization, but rather the regulation of the enzyme activity, that is of importance for its function (Kvamme and Olsen, 1980). The importance of glutamine as a precursor for transmitter glutamate and GABA is well recognized and will be discussed below.

Glutamate dehydrogenase (E.C. 1.4.1.2.) has been localized to nonsynaptic mitochondria (Reijnierse et al., 1975) by subcellular fractionation studies. The enzyme activity has been found to be higher in an astroglial fraction than in Purkinje cell or granule cell fractions from cerebellum (Patel, 1982). After infusion with [^{15}N]–ammonium acetate, the labeling of the α-amino group of glutamine is tenfold higher than that of glutamate. This is consistent with a localization of the enzyme in the small compartment. Evidence therefore favors a preferential localization of the enzyme in the astroglial cells.

Pyruvate carboxylase catalyzes the formation of oxaloacetate from pyruvate and leads to a net synthesis of a TCA cycle constituent. The enzyme has been shown by immunohistochemical techniques to be highly concentrated in the astrocytes (Shank et al., 1981). It has also been found in high concentrations in a primary culture of astrocytes (Yu et al., 1983).

The results of an extensive investigation into several energy-metabolizing enzymes in free and synaptic mitochondria are summarized elsewhere (Leong et al., 1984). The localization of GABA aminotransferase (GABA-T) is discussed below.

Information on the localization of the glutamate enzymes is summarized in Fig. 1.

The kinetic studies on the labeling of glutamate, glutamine, and GABA from the different precursors have led to the following simplified model (Fig. 2). Glucose and its related metabolites are the precursors of the large glutamate pool (Table 1). The large glutamate pool is in the neuron, where the synthesis of the transmitter glutamate, aspartate, and GABA takes place. After the transmitters have been released from the nerve terminals, they are partially taken up into the nerve terminals or into the glial cells, where they are converted to glutamine, as discussed above. The transmitter amino acids therefore transport the carbon skeleton from the neurons to glial cells (van den Berg et al., 1975). On the other hand, glutamine diffuses freely between the cells and is believed to transport the carbon skeleton from the glial cell to the

LARGE COMPARTMENT			SMALL COMPARTMENT
GAD			GNS
GABA-T	gaba-T	gaba-T	GABA-T
gdh	gdh	gdh	GDH
GLNase	GLNase	glnase	glnase
ASP-T	ASP-T	asp-t	asp-t
			PC
GABA NEURONS	Glu NEURONS	OTHER NEURONS	ASTROCYTES

Fig. 1. The distribution of enzymes between the large and small glutamate compartment. The large compartment has been separated into GABA neurons, Glu neurons, and other neurons. The small compartment in this context represents the astrocytes. Abbreviations: GAD, glutamate decarboxylase; GABA-T, GABA aminotransferase; GDH, glutamate dehydrogenase; GLNase, phosphate activated glutaminase; Asp-T, aspartate aminotransferase; PC, pyruvate carboxylase; GNS, glutamine synthetase.

neuron. The model has been extensively reviewed and discussed in several papers (van den Berg et al., 1973, 1975, 1978). A large body of literature supports the view that glutamine is an important precursor for transmitter glutamate and GABA. In the synaptosome fraction (Bradford et al., 1978) and hippocampal slices (Hamberger et al., 1979a,b), glutamine has been shown to be the best precursor for transmitter glutamate. In vivo experiments have shown that glutamine can function at least as well as glucose as a precursor of transmitter glutamate or GABA (Berl et al., 1961; Thanki et al., 1983; Ward et al., 1983). In addition, during hypoglycemia glutamine is almost maximally reduced before glutamate is reduced to a significant extent (Engelsen and Fonnum, 1983). This is also consistent with glutamine being an important precursor for glutamate. Acetate and other precursors of group 1 preferentially label the small glutamate compartment. In agreement with these data, fluoroacetate inhibits the formation of glutamine (Berl and Clarke, 1978). Citrate and fluorocitrate seem to be even more restricted in labeling or inhibiting the small glutamate compartment (Cheng, 1972).

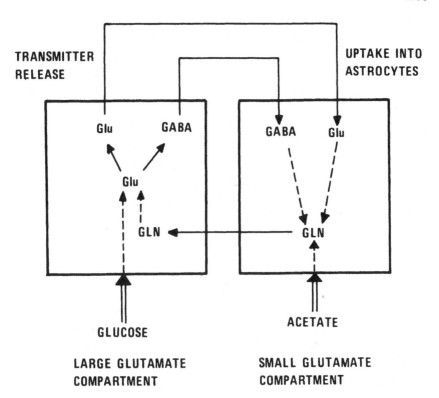

Fig. 2. The simple two-compartment model in brain. The two compartments are connected through the flow of glutamine from the small to the large compartment and by the transport of glutamate and GABA from the large to the small compartment.

4. Turnover of Glutamate

4.1. Localization of Glutamate Pools in the Brain

Glutamate has a central role in brain metabolism, as well as being a transmitter and a precursor for GABA. It is therefore of interest to define and quantify the different glutamate pools. The size of these pools will differ from area to area and species to species (Fig. 3).

The transmitter pool of glutamate may be quantified after selective degeneration of the glutamate terminals by surgical or chemical lesions. By this technique the transmitter glutamate pool has been found to be 20–30% of the total glutamate content in neostriatum, thalamus, lateral septum, nucleus accumbens, and

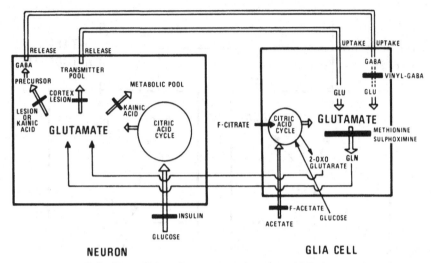

Fig. 3.   Glutamate cycles in the brain. The scheme shows the roles of glutamate in the brain. In the neuron, glutamate may act as a transmitter, precursor for GABA, or in the metabolic pool. An example is shown of how the three pools can be selectively destroyed in neostriatum. Glutamate is formed from glucose through the citric acid (TCA) cycle, from glutamine or 2-oxoglutarate. The latter two are derived from the glial cells.

Within the glial cell, glutamate is partly formed through the TCA cycle, and partly from the glial uptake of the transmitters glutamate and GABA. Glutamate is further metabolized to glutamine. Moreover, the figure shows that glutamate formation via the TCA cycle may be inhibited by fluoracetate and fluorocitrate. The formation of glutamate from GABA may be inhibited by GABA-T inhibitors, e.g., vinyl-GABA. Finally, the formation of glutamine may be inhibited by methionine sulfoximine.

fornix terminal regions (Fonnum et al., 1981; Walaas and Fonnum, 1980; Lund Karlsen and Fonnum, 1978).

Similarly the pool of glutamate as a precursor for GABA can be estimated after selective destruction of GABAergic terminals. In substantia nigra, the region with the highest content of GABAergic terminals, selective destruction of GABA terminals is accompanied by small glutamate changes (Korf and Venema, 1983; Minchin and Fonnum, 1979). The precursor pool of glutamate is therefore probably less than 10% of the total glutamate pool in any brain region.

The proportion of glutamate in the small glutamate pool, i.e., the glial cells, can be estimated from the relative specific radioactivity of glutamine compared to glutamate after administration of group 1 precursors (acetate, aspartate, $NaHCO_3$). In several regions of cat brain this ratio is 5, which indicates that the glial pool is 20% of the total glutamate pool (Berl et al., 1961). As similar reduction in glutamate is obtained after administration of fluoroacetate, which presumably inhibits the TCA cycle in the glial pool (Fonnum et al., to be published). In this respect, it should be borne in mind that acetate may not be exclusively localized to the small pool and that the figure is possibly an overestimation. Based on kinetic models, Cremer et al. (1975) suggested that the glial pool of glutamate was very small, perhaps in the order of 5%.

The remaining glutamate must be ascribed to other neuronal structures and coupled to the general metabolism in neurons. This is by far the largest pool and probably accounts for about half of the glutamate content in the brain. In neostriatum, the neuronal cell bodies are destroyed by kainic acid injection. Such treatment is accompanied by 50% loss of glutamate in the neostriatum (Nicklas et al., 1979).

4.2. Turnover Studies

Since transmitter glutamate is synthesized in the large glutamate pool, group 2 precursors (glucose, and so on) should be used for studying the turnover of transmitter glutamate. Regions with high transmitter glutamate content are cortex, striatum, and hippocampus; regions with low transmitter glutamate content are globus pallidus, substantia nigra, and medulla (Fonnum, 1984). There is a correlation to a certain extent between the transmitter glutamate content and glucose metabolism in several brain regions. This could be taken as evidence for good correlation between glucose and glutamate transmitter metabolism (Hawkins et al., 1983). At present it is not known whether this is of importance or only serendipitous.

There are several different types of glucose isotopes that may be used for studying glutamate turnover. Incorporation of (1-^{14}C)-glucose is much more rapid and extensive than incorporation from (2-^{14}C)-glucose since the former labels the methyl carbon and the latter, the carboxyl-carbon in acetylCoA. For the same reasons, (2-^{14}C)-acetate gives more extensive incorporation than (1-^{14}C)-acetate. On the other hand, the carboxyl carbon will be lost more rapidly on the next turn of the Krebs cycle and therefore diminish the problem of recycling of the label. (3,4-^{14}C)-Glucose gives labeling of the carboxyl group in pyruvic acid that is rapidly lost, and gives only 10% of the labeling obtained with [1-^{14}C]-glucose (van den Berg, 1973).

Pulse labeling of glutamate with glucose shows a reasonably good precursor–product relationship (van den Berg, 1973). Although there are several metabolites between glucose and glutamate, the incorporation from $[2\text{-}^{14}C]$-glucose is such that the labeling of glutamate seems to be directly correlated with glucose. This probably means that the rate of conversion of 2-oxoglutarate to glutamate is much faster than the conversion of 2-oxoglutarate to succinate.

In studies of the incorporation of label into glutamate it is therefore essential to have a picture of the flow of glucose. Treatment of animals with di-*n*-propylacetate gives a decrease in glutamate incorporation that may be traced at least in part to the decrease in the incorporation of glucose (Cremer et al., 1978). Similarly, in 6-aminonicotinamide-treated animals there is a decrease in both glucose utilization and the incorporation of label into glutamate and GABA from $[U\text{-}^{14}C]$-glucose (Gaitonde et al., 1983). In these studies the decrease in glutamate labeling was therefore not necessarily a consequence of the low turnover rate of glutamate, but a decrease in glucose metabolism. Infusion of labeled glucose intravenously for 10 min gave similar incorporation of label in glutamate in regions with high and low transmitter glutamate content (Bertilsson et al., 1977). In this study, designed to assess the turnover of GABA, no data on the glucose flux was given. The results could indicate that the metabolic pool of glutamate dominated and that the transmitter pool was therefore hidden.

There are studies, however, that tend to indicate a specific change in the incorporation of label into glutamate. In one study, rats were left in a state of deep anesthesia with pentobarbital. In these animals there was a significant decrease in the specific activity of glutamate, aspartate, and glutamine, but not in total radioactivity, in the brain 3 min after giving $[2\text{-}^{14}C]$-glucose intravenously. Similarly, 2 h after triethyltin treatment there was a significant decrease in the label incorporated into the same amino acids compared to the incorporation of total radioactivity in the tissue. Ten minutes after giving $[2\text{-}^{14}C]$-glucose, the difference between incorporation into the amino acids and the total activity in the brain was diminished (Cremer and Lucas, 1971). In another type of experiment, ammonium acetate and portocaval anastomosis increased the release of glutamate in the cortex cup. Interestingly, the same two conditions also increased the incorporation of label in glutamate after 10 min infusion with glucose. In this case, there was unfortunately no data on the glucose flux (Moroni et al., 1983). Therefore, the results from glucose pulse or glucose infusion indicate that this is not a straightforward method for determining glutamate turnover. It may be, however, that pulse labeling and short time are essential features for obtaining data on the turnover of transmitter glutamate.

Glutamine is another precursor that may be used for measuring the rate of incorporation of label into glutamate. In slices and synaptosomes, glutamine is a several-fold more efficient precursor than glucose for transmitter glutamate (Hamberger et al., 1979a,b; Bradford et al., 1978; Reubi, 1980). The incorporation of label from glutamine after intracerebral injection is, however, very slow. The specific activity of the glutamine pool is 10–50 times higher than that of glutamate (Costa et al., 1979; Thanki et al., 1983; Berl et al., 1961). This makes the use of glutamine as a precursor less attractive for practical work.

Labeled glutamine may also be obtained from labeled acetate, and the turnover of glutamate could then be calculated from glutamine. Labeled acetate gives a good incorporation of label into glutamate. But the pool of glutamine for glutamate synthesis should be derived from the released glutamate and GABA, in addition to the TCA cycle in glial cells. Therefore, acetate can only partly label the precursor glutamine pool. An additional complication is that glutamine is also the precursor of the pool of glutamate used in GABA synthesis. It is of interest in this context to recall that neither pentobarbital treatment nor triethyltin treatment were accompanied by any changes in glutamate labeling compared to radioactivity in the total tissue from [1-^{14}C]-acetate or [1-^{14}C]-butyrate. There were small decreases in the incorporation of label in glutamate compared to glutamine after only 10 min (Cremer and Lucas, 1971).

As early as 1973 van den Berg suggested that oxoglutarate from the glial cells could constitute an important precursor for glutamate in the large pool. 2-Oxoglutarate has recently been advanced as a precursor for transmitter glutamate in the brain (Shank and Campbell, 1982, 1984). The 2-oxoglutarate is formed in the astroglia from glucose through pyruvate carboxylase and transferred from the glial cell to the neuron. Shank and Campbell (1984) have shown that there is a high-affinity uptake of 2-oxoglutarate into the brain synaptosomes, and we have recently shown that it is at least in part specific for glutamergic terminals (Fonnum et al., unpublished). The rate of uptake is, however, probably too low to cope with the synthesis of transmitter glutamate. So far, few data exist to support the idea that ornithine may be an important precursor for glutamate (Yoneda et al., 1982).

There are several examples of pharmacological manipulation leading to a change in the level of the transmitter amino acids. Such changes may in certain cases be caused by an increase in turnover rate, which exceeds the synthesizing capacity of the tissue. Electrical kindling gives a small decrease in cortical glutamine level, possibly owing to the increased turnover of glutamate (Leach et al., 1983). Hypoglycemia leads to a decrease in glutamine and glutamate and an increase in aspartate (Butterworth et al., 1982). These changes may be caused by a decrease in the supply of glucose so that the tissue cannot maintain the rate of glutamate production. Thus the amino acid changes were higher in a control neostriatum than in a decorticated neostriatum, possibly because of the higher turnover in the former (Engelsen and Fonnum, 1983). Treatment of fasted rodents with 2-amino-7-phosphonoheptanoic acid, a glutamate/aspartate antagonist, is accompanied by a decrease in aspartate and glutamate levels in neostriatum. The decrease could be caused by an excess release of the two transmitter amino acids (Westerberg et al., 1983).

5. Turnover of GABA

5.1. Inhibition of GABA Metabolism

GABA is synthesized in brain almost exclusively from glutamate by GAD (Baxter, 1976). The level of GAD and, to some extent, the availability of the coenzyme pyridoxal phosphate are the limiting factors for GABA synthesis. Recently, a scheme was suggested showing how the enzyme activity, or its binding to pyridoxal phosphate, could be regulated by ATP, phosphate ion, and GABA (Porter and Martin, 1984). There has been some discussion whether cysteic acid decarboxylase and GAD are similar or different enzymes. There exists a cysteic acid decarboxylase with a high affinity for cysteic acid, and no affinity for glutamate. This enzyme, which will decarboxylate cysteine sulfinic acid, is believed to be exclusively responsible for the synthesis of taurine. GAD itself has a low affinity for cysteic acid and glutamate is a competitive inhibitor of this activity. GAD activity is therefore believed to be solely responsible for the GABA synthesis and not important for taurine synthesis (Wu, 1982).

GABA is metabolized by GABA-T to succinic semialdehyde, which is further oxidized by succinic semialdehyde dehydrogenase to succinic acid. Inhibition of GABA-T is therefore accompanied by an increase in GABA and this increase should reflect GABA turnover. The inhibition of GABA-T can be achieved either by the use of specific inhibitors or by studying the postmortal increase of GABA.

5.2. Use of GABA-T Inhibitors

Experiments in our laboratory have shown that about 50% of the GABA-T activity must be inhibited to give an adequate increase in GABA level in the brain (Omholt-Jensen, 1984). There are three basic conditions that must be fulfilled if GABA-T inhibitors are used to study GABA turnover.

First, the rate of synthesis of GABA must not be changed during the inhibition. Many of the GABA-T inhibitors interact with pyridoxal phosphate, which is not only a coenzyme for GABA-T, but also for GAD, AspT, alanine aminotransferase, and ornithine decarboxylase (Metcalf, 1979). Several of the GABA-T inhibitors therefore inhibit other enzymes, in particular GAD, and cannot be used in turnover studies. For some inhibitors, such as amino oxyacetic acid (AOAA), the inhibition of GAD is species dependent (van Gelder, 1966). There has been no report showing that GABA synthesis in vivo can be regulated by changes in the substrate pool of glutamate.

Second, an increased level of GABA should not inhibit the synthesis of GABA. We know, however, very little about the regulation of GABA synthesis. GABA alone is a poor inhibitor of GAD (Wu, 1982). There are several indications that the enzyme activity may be regulated through its association with pyrodoxal phosphate (Miller et al., 1978). We did not find any difference in the accumulation of labeled GABA from labeled glucose and acetate either 30 min or 5 h after administration of the GABA-T inhibitor AOAA (Minchin and Fonnum, unpublished). In agreement, Chapman et al. (1982) did not find any changes in the rate of formation of labeled GABA in cortex between γ-vinyl-GABA-treated rats and control rats. Since several studies show a linear increase of GABA with time after administration of GABA-T inhibitors, it may be assumed that the increase in GABA level does not regulate the GABA synthesis (Jung et al., 1977a; Walters et al., 1978; Ionesco and Gale, 1979). In this respect, it has been generally accepted that GABA synthesis and metabolism take place in different cellular compartments, namely the nerve terminal and the glial cells, respectively. Subcellular fractionation studies and lesion studies of GABAergic fibers accord with such localization (van Kempen et al., 1965; Kataoka et al., 1974). Subcellular fractionation of GABA pools after inhibition of GABA-T indicates a higher increase of GABA in the nonsynaptosomal cytoplasm than in the synaptosome fraction (Geddes and Wood, 1984; Pagliusi et al., 1983). An exception is the study with γ-vinyl-GABA by Sarhan and Seiler (1979). Recent studies with a pharmacohisto-chemical method for GABA-T have, however, shown the enzyme to be present in GABA neurons with a particularly rapid recovery in the GABAergic nerve terminals after inhibition with AOAA (Vincent et al., 1980). I therefore conclude that the highest pro-portion of GABA-T activity is present in glial cells and in the GABAergic cell bodies. The presence of GABA-T in the GABAergic nerve terminal cannot be questioned, but the propor-tion of the enzyme may be relatively low. It follows from most of these studies that the GABA increase is not in the same compart-ment as GABA synthesis, and that the increase therefore has little impact on the rate of GABA synthesis.

Third, GABA inhibitors should not impose serious physio-logical changes in the animal. During GABA-T inhibition, physio-logical changes, such as sedative, anticonvulsive, antinociceptive, and antianxiety effects, have been reported (Palfreyman et al., 1981). Such effects are clear indications that the increase of GABA may influence the transmission of other synaptic systems in the brain and therefore indirectly cause changes in metabolism and turnover of other neurotransmitters. The physiological changes themselves do not necessarily affect GABA turnover, although this is a factor that should be borne in mind. Hypothermia has been observed with GABA-T inhibitors and may result in a de-crease in both GABA synthesis and metabolism (Löscher and Vetter, 1984).

The physiological effects that are unwanted in turnover stud-ies may be modified by using lower doses of GABA-T inhibitors or by applying the inhibitors intracerebrally to avoid systemic effects.

5.3. Properties of Different GABA-T Inhibitors

Aminooxyacetic acid has been studied extensively as an inhibitor of GABA-T since it was first introduced by Wallach (1961). Since AOAA reacts with pyridoxal phosphate, it inhibits several other enzymes in higher doses. In mouse brain, the accumulation of GABA is linear for several hours after administration of 20–40 mg AOAA/kg body wt (van Gelder, 1966). In rat brain, similar studies yield slightly varying results. Collins (1972) found an increase of GABA with both 25 and 40 mg AOAA/kg body wt, and the time course for the higher concentration was linear for 2–3 h. Walters et al. (1978) investigated the regional increase of GABA with sev-eral AOAA concentrations and found 25 mg/kg to be an accepta-ble dose for such studies. This dose gave 80–90% inhibition of

GABA-T in most regions of the brain except substantia nigra, which showed only 43% inhibition. Walters et al. (1978) found large regional variations in turnover time for GABA, whereas Bernasconi et al. (1982) found similar turnover time in cortex, striatum, hippocampus, and cerebellum after 20 mg AOAA/kg. Yamatsu et al. (1982) found considerable regional variation in turnover time with 50 mg AOAA/kg. Löscher and Vetter (1984) also found considerable regional differences after giving 30 mg AOAA/kg. Ionesco and Gale (1979) have used the method to study GABA turnover in normal and deafferentiated substantia nigra.

γ-Vinyl-GABA is another inhibitor that has been used extensively for GABA-turnover studies. Similar to AOAA, γ-vinyl-GABA has a rapid onset of inhibition. When administered systemically, high doses (800–1500 mg/kg body wt) should be used. Such doses inhibit GABA-T 80–90% and have only a small effect upon GAD. The compound has no effect on aspartate or alanine aminotransferase (Jung et al., 1977a). γ-Vinyl-GABA has no effect on the receptor binding of GABA, but is said to have some effect on GABA uptake in vitro (Löscher, 1980). The increase in GABA is linear for 2–4 h after systemic injections of 1000 mg/kg (Jung et al., 1977a). This dose of γ-vinyl-GABA does not alter the rate of GABA synthesis from glucose (Chapman et al., 1982). In contrast to the results with AOAA, γ-vinyl-GABA has been found to increase synaptosomal GABA more than other GABA (presumably glial) pools (Sarhan and Seiler, 1979).

In small doses (10–50 μg/μL), γ-vinyl-GABA has been administrated stereotaxically into the brain for studying turnover in restricted regions. The animals should be anesthetized with short-lasting anesthesia, such as ether or halothane, to avoid prolonged effects on the general metabolism of the brain. GABA-T is inhibited more than 80% and the increase of GABA is linear for almost 3–4 h (Casu and Gale, 1981; Omholt-Jensen, 1984). Under these conditions, GAD is inhibited less than 10% (Casu and Gale, 1981). By injection of 10 μg/μL γ-vinyl-GABA, the opposite side is inhibited less than 20% and may be used as an uninjected control. γ-Vinyl-GABA is an amino acid analog that may be measured in the sample in the same way as GABA, both with HPLC and "classical" amino acid analysis with fluorescence detection of its *o*-phthaldehyde derivative (Omholt-Jensen, 1984). It is therefore possible to correlate the increase in GABA with the γ-vinyl-GABA content in the sample and thereby have an internal control of the injection. With this procedure it is possible to detect large variations in the turnover time for GABA from region to region (Casu and Gale, 1981). It has been used to study the effect of various drugs and lesions on the GABA turnover in superior colliculus (Melis and Gale, 1983). γ-Vinyl-GABA is not commercially available, but may be obtained on request from Merrel International, Strasbourg.

Gabaculine is another irreversible inhibitor of GABA-T that has been used for turnover studies. The compound has no effect on either GAD or aspartate and alanine aminotransferase (Matsui and Deguchi, 1977). Gabaculine has no effect on receptor binding, but a weak effect on uptake (Löscher, 1980). A slow onset of action occurs when administered systemically, and a full inhibitory effect is not obtained before 15–30 min after administration (Rando and Bangerter, 1977). The increase of GABA is linear for 15–75 min (Bernasconi et al., 1982). When gabaculine is administered systemically, maximal effect on GABA increase is found with a dose of 100 mg/kg body wt intraperitoneally.

Gabaculine may also be injected intracerebrally, and 100 μg/2 μL in striatum inhibits GABA-T more than 90% at 15–30 min after the administration. The increase in GABA is linear for the first hour after administration (Giorgi and Meek, 1984). This procedure has been used to investigate the effect of apomorphine, kainic acid, and several cholinergic drugs on GABA turnover in striatum and median raphe nuclei (Forchetti et al., 1982).

Ethanolamine-O-sulfate (EOS) penetrates the blood–brain barrier poorly and is generally administered intracerebrally (Fowler and John, 1972). EOS has no inhibitory effect in vitro toward GAD or alanine and aspartate aminotransferase (Fowler and John, 1972). It does not have any effect on GABA uptake, but has a weak effect on the receptor binding (Löscher, 1980). High doses, when given systemically, will inhibit GABA-T in brain (Fletcher and Fowler, 1980). Intraventricular injection of EOS has been used to study GABA turnover in striatum and substantia nigra (Cattabeni et al., 1979).

γ-Glutamyl hydrazide (γ-GH) reacts with pyridoxal phosphate and therefore inhibits GAD as well as GABA-T (Massieu et al., 1964). It is thus surprising that this compound has been used to study GABA turnover in brain. Interestingly, Perez de la Mora et al. (1977) observed an increase in the turnover of GABA with this inhibitor after stimulation of the striato-nigral tract, and a decrease in turnover 3 h after interruption of the tract. These are important criteria for showing that the changes found are caused by turnover. γ-GH used at doses of 160 mg/kg, to reduce inhibition of GAD, produces only small increases in GABA (Fuxe et al., 1979). γ-GH has a slow onset of inhibition, and the increase in

GABA after systemic injection is relatively smaller after 1 h than after 3 h (Walters et al., 1978). The method has been used to show a reduced turnover of GABA in striatum and substantia nigra after apomorphine administration (Hökfelt et al., 1976).

There are several other GABA-T inhibitors that are not considered suitable for turnover studies or have not been extensively studied. γ-Acetylenic-GABA is an irreversible GABA-T inhibitor that is active at a much lower dose (100 mg/kg ip) than γ-vinyl-GABA. The inhibitor exhibits marked inhibition of GAD, however, and is therefore considered less suitable (Jung et al., 1977b). It has been applied by Fuxe et al. (1979) to study the effect of chloropromazine and sulpiride and the effects of several cholinergic drugs (oxotremorine, physotigmine) on the turnover of GABA.

L-Cycloserine is an inhibitor of several aminotransferases, including GABA-T. It interacts with pyridoxal phosphate and therefore inhibits GAD, although to a lesser extent than GABA-T (Dann and Carter, 1964).

In conclusion, γ-vinyl-GABA appears to be the most suitable inhibitor to be used in GABA turnover studies. Gabaculine is another strong candidate whose only disadvantage is its slow onset of inhibition. This disadvantage may, however, not be of practical importance. Ethanolamine-O-sulfate has not been extensively studied, probably because of its lack of availability. It could be classified similarly to gabaculine. AOAA has been widely studied and seems to be a very suitable compound, at least for mouse brain.

5.4. Postmortal Changes in GABA

It is well recognized that there is a rapid postmortal increase of GABA in the brain (Minard and Mushahwar, 1966). The increase is caused by GABA synthesis proceeding initially at the same rate postmortem as in vivo, and an almost complete inhibition of GABA catabolism by anoxia. The latter occurs because GABA transamination requires 2-oxoglutarate for transamination and oxidized NAD for the oxidation of succinic semialdehyde.

The increase of GABA has been found to be maximal during the first 4 min postmortem, but it proceeds with a high rate for more than 30 min (Patel et al., 1970, 1974). When [U-^{14}C]-glucose is used as a precursor, there is a dramatic increase in the specific activity of GABA compared to glutamate from 0.6 to 1.0 during the first 4 min postmortem. During the following 30 min, the ratio remains constant. This is further evidence that the glutamate pools in the brain are heterogeneous, and is also an indication of a small, highly labeled pool of glutamate that is coupled to GABA synthesis (Patel et al., 1974). The postmortal increase in GABA could be used to study GABA turnover in brain. This would involve the advantage of not employing a GABA-T inhibitor. The success of the method would depend on three factors: (1) GABA synthesis postmortem reflects the synthesis in vivo, (2) the GABA metabolism is fully inhibited postmortem, and (3) the regulation of GABA synthesis in vivo is operating postmortem.

There is evidence in the literature that GAD has a higher degree of binding to pyridoxal phosphate postmortem (Miller et al., 1978). This could imply that the GABA synthesis postmortem is higher than in vivo. I cannot believe, however, that such an effect can be of great importance. The postmortal increase of GABA 10 min after decapitation is fairly well correlated to the GAD activity in different brain regions (Yamatsu et al., 1982). The postmortal increase of GABA is decreased by GAD inhibitors, either methoxypyridoxine administrated 1 h before or mercaptopropionic acid 2 min before decapitation. We can, therefore, conclude that GABA synthesis postmortem proceeds as in vivo. The decrease in GABA synthesis observed during a longer incubation period post mortem may be caused by the inactivation of GAD. But it would be very serious for turnover studies if this decrease in synthesis reflected a change in GABA synthesis between the different compartments. The problem could be circumvented by selecting short observation times postmortem.

Adminstration of AOAA (50 mg/kg) 1 h before decapitation increases the postmortal rise in GABA by as much as 50% (Yamatsu et al., 1982). There are substantial regional variations from no effect of AOAA on the increase of GABA postmortem in substantia nigra and frontal cortex, to large effects in hippocampus and superior colliculus. Therefore, GABA metabolism does not seem to be completely or immediately inhibited by anoxia.

It is more difficult to evaluate whether the in vivo regulation of GABA synthesis also operates postmortem. In general, the increase in GABA found for the whole brain during the first 5 min postmortem corresponds with the values obtained after extensive GABA-T inhibition (Patel et al., 1974; Jung et al., 1977a). However, in comparing the increase of GABA postmortem with that obtained by 50 mg/kg AOAA in different regions, there is a negative correlation (Yamatsu et al., 1982). Thus, substantia nigra shows the highest postmortal increase, but the lowest increase with several GABA-T inhibitors (Yamatsu et al., 1982; Walters et al., 1978; Melis and Gale, 1983).

Two papers have elucidated in some detail the turnover of GABA under different experimental conditions based on postmortal increase. Mansky et al. (1981, 1983) investigated the involvement of GABA turnover on endocrine release. They found a linear increase of GABA between 1 and 4 min postmortem in several brain structures. They minimized errors by using the two halves of the brain for the two time points. The short time intervals also ensured that they were working within the time interval, in which the highly labeled glutamate pool was still present. To my mind this procedure and concept is excellent, but the GABA level increased only 5–15% during the three minutes. Considering experimental difficulties, it is obvious that the increase is too low to give a high degree of reproducibility.

The authors claim to find significant changes in GABA turnover with this method (Mansky et al., 1981, 1983). Pericic et al. (1978) compared the effect of α-butyrolactone, a drug inhibiting GABA turnover, and picrotoxin, a drug accelerating GABA turnover, on several cortical structures. The animals were decapitated and the brains incubated at room temperature for 5 min to study the postmortal increase, which ranged from 10 to 50%. Although the results pointed in the expected direction, the differences were too small to be significant. The effects of the two drugs on turnover were clearly seen from the increases of GABA obtained after AOAA treatment.

5.5. Turnover of GABA by Infusion with Labeled Precursors

Glutamate is the immediate precursor for GABA in the brain. This acidic amino acid is, however, only poorly transported across the blood–brain barrier and cannot be administered systemically. When labeled glutamate is administered intracerebrally, it is rapidly taken up into the glial cells (Schon and Kelly, 1974) and converted to glutamine. The conversion of glutamate into GABA based on the relative specific activity is poor (0.1–0.3) (Berl et al., 1961).

Glutamine, in contrast to general belief, is poorly transported into the brain. When injected intraventricularly, it will give rise to a high relative specific activity of GABA compared to glutamate (0.5–0.9) (Berl et al., 1961). The rate of conversion of labeled glutamine into glutamate and GABA (expressed as cpm/mol) is very poor compared to other precursors. Such results were obtained by Costa et al. (1979) after intraventricular injection of glutamine, and by Thanki et al. (1983) after perfusing the sensorimotor cortex with glutamine. The low rate of incorporation may be caused in part by the efficient dilution of labeled glutamine with glutamine stored in vivo. On the other hand, Gauchy et al. (1980) were able to study the release of [^{3}H]-GABA after continuous infusion of [^{3}H]-glutamine.

Other studies have taken advantage of the fact that glucose will rapidly label the large pool of glutamate, which, at least in part is responsible for GABA synthesis. In these studies, labeled glucose may be given as a single pulse or by continuous infusion. The incorporation of label into glutamate and GABA has been monitored after glucose was infused intravenously for 10 min (Bertilsson et al., 1977). The fractional rate constant for the efflux of labeled GABA was calculated from the isotopic enrichment of glutamate and GABA. Thus an apparent increase or decrease of the turnover could be caused by the changes in the incorporation of label into either glutamate or GABA. The calculation was based on the assumption that the precursor glutamate compartment is open and in rapid equilibration with the rest of the glutamate in the tissue. Thus the equations originally developed for acetylcholine turnover from choline infusion were used for the calculations (Racagni et al., 1974). The procedure is also similar to that developed for measuring the turnover of biogenic amines (Neff et al., 1971).

There are several reasons why the use of glucose as a precursor for GABA must be treated with some skepticism. There is no correlation between the rate of glucose turnover and the level of GABA in different regions. Thus glucose turnover is high in several cortical regions and low in substantia nigra and globus pallidus (Hawkins et al., 1983). As previously discussed, the rate of glucose utilization in brain is about 0.8 μmol/min/g wet wt, which results in a rate of acetylCoA formation equal to 1.6 μmol/min/g wet wt (Hawkins et al., 1983). Cremer et al. (1978) and Bertilsson et al. (1977) arrived at a rate of GABA turnover that is similar to, or even greater than, the rate of glucose metabolism. This is obviously too high a rate of GABA synthesis. The authors based their estimation on the assumption that GABA was formed from the total glutamate pool.

How large a part of the glutamate pool is available for GABA synthesis is an open question. Most studies suggest that only a small part of glutamate serves this function. Postmortem studies on the changes in the specific activity of glutamate and GABA suggest that only 2% of the glutamate pool is responsible for 80% of the GABA synthesis (Patel et al., 1974). Degeneration of the GABAergic terminals in the substantia nigra after interruption of the striatal nigral tract suggest that although 8 μmol of GABA was lost, only 1 μmol of glutamate was lost (Korf and Venema, 1983; Minchin and Fonnum, 1979).

If we assume that the precursor glutamate pool has a higher specific activity than the total glutamate pool, as indicated by postmortem studies, then the calculated rate of GABA synthesis from glutamate could be reduced correspondingly.

Studies on labeled glucose have revealed that only in substantia nigra does the enrichment of label in glutamate and GABA reflect the expected precursor–product relationship (Bertilsson et al., 1977). In other regions, such as neostriatum, nucleus accumbens, and globus pallidus, the results were close to the expected precursor–product relationship, whereas in cortical structures there was no correlation. In cortical structures, a high percentage of newly synthesized Glu may play a transmitter role and therefore occlude the results. Thus the method should be limited to studying GABA turnover only in brain nuclei with a high GABA content.

When turnover rates were calculated for the first 7 min (Bertilsson et al., 1977), the fractional rate constant for GABA decreased. This may be attributed to the progressive recycling of radiolabel through the GABA cycle. Errors may be minimized by reducing the period for calculating the fractional rate constant. Even if the method does not give correct turnover for GABA, it has proved useful for comparative studies and as such has been of considerable interest. No other single method gives so much data on turnover studies as this procedure (Marco et al., 1978; Costa et al., 1978; Moroni et al., 1978; Revuelta et al., 1981). The fact that there is an increase in turnover of GABA in the substantia nigra (Mao et al., 1978) following stimulation of the striatonigral tract has provided considerable support for this method.

In pulse labeling experiments with [2-^{14}C]–glucose, both Chapman et al. (1982) and Cremer et al. (1978) showed a decrease in labeling of GABA, but not in glutamate or glucose, during valproate treatment. This indicates a change in turnover of GABA.

Errors caused by recycling are expected to be even larger when studying the decrease in specific activity of GABA 60–90 min after a pulse with [U-C^{14}]–glucose, as suggested by Smith et al. (1982). This method is also based on the assumption that the GABA is present in an open compartment. The authors assume that a dominating proportion of the amino acid synthesis, at least in forebrain, is linked to neurotransmitter function. Neither of these two assumptions are correct.

The half–life was calculated from a first–order reaction decay rate and the turnover was the reaction rate constant multiplied by the amino acid level.

6. Turnover of Glycine

Glycine probably has a transmitter function in the medulla and spinal cord grey, but not in other brain regions. The glycine level is 3–5 times higher in medulla and spinal cord than in other regions of the brain. It is therefore quite natural to compare the biosynthesis of glycine or glycine flux in these regions with that of other brain regions (Shank and Aprison, 1970).

Glycine can be supplied to the brain from three main sources: exchange or flux from the blood and biosynthesis via serine or glyoxalate (Fig. 4) (Aprison and Nadi, 1978). The exchange or flux from the blood is slow compared to the biosynthesis. The flux is also independent on the glycine level in the brain region and is highest in the cerebellum (0.149 μmol/g/h) and lowest in the spinal cord grey (0.031 μmol/g/h) (Shank and Aprison, 1970).

The main path of biosynthesis of glycine is believed to go through serine for several reasons. Labeling with $(3,4-^{14}C)$-

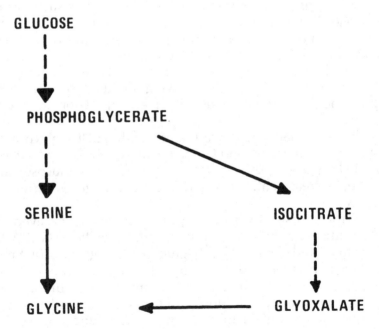

Fig. 4.   Synthesis of glycine.

glucose will label the carboxyl group in serine and glycine, whereas (1-^{14}C)–glucose will label glyoxylate and 2-carbon atom in serine. The latter is lost on conversion of serine to glycine. The labeling in glycine is five times higher with (3,4-^{14}C)– than with (1-^{14}C)–glucose, indicating that serine is the most important precursor. The specific activity ratio between glycine and serine was similar in most regions when either serine or glucose was used as the precursor (Shank and Aprison, 1970; Shank et al., 1973). The synthesis of serine from glucose seems to be much higher than the flux from blood (0.1–0.2 μmol/g/h) (Shank and Aprison, 1970).

The rate of formation of serine and glycine from glucose has been estimated from the simple equation:

$$F = C(S_b/\Delta t)/(S_a - S_b)$$

where C is the concentration of the product, and S_a and S_b are the average specific activities of the precursor and the product, respectively, during the time interval Δt. The estimation is also based on the assumption that serine and glycine synthesis are not compartmentalized, and it does not take into account the metabolism of the product. Using this approach, Shank and Aprison (1970) have provided the only data available on glycine turnover. The rate of formation of glycine from serine is about 0.8 μmol/g/h and the formation of serine from glucose about 2 μmol/g/h. There are no significant differences between the regions with high or low concentrations of glycine. The slow rate in the transmitter-rich regions could be taken as an indication that reuptake and reuse of glycine in the nerve terminals are very efficient.

The turnover rates of glycine in different regions have also been calculated from the assumption that the [14C]-amino acids disappear from one open pool at steady-state conditions (Freeman et al., 1983). The decrement in radioactivies at 60 and 90 min postinjection of the glucose pulse has been employed to calculate the half-life.

The turnover rate for glycine has been found to be 0.2–1.0 μmol/g/h (Freeman et al., 1983). The rate was low in striatum and cortex, and high in the cerebellar granular layer. The method does not take into account recycling of labeling.

7. Conclusions

There are four amino acids (aspartate, GABA, glutamate, and glycine) that are strong transmitter candidates in brain. The difficulties in developing methods for studying their turnover are partly owing to the complex compartmentation of their metabolism in general. The transmitter pool is often difficult to separate from the other functional pools, and there is a multitude of different synthetic and metabolic pathways for each transmitter amino acid.

There exists a large body of literature on the turnover of GABA in brain. In this review I have recommended the use of specific GABA-T inhibitors, such as γ-vinyl–GABA, gabaculine, and aminooxyacetic acid, that can be administered either systemically or intracerebrally.

A method based on the use of glucose infusion has been widely used. The method is based on the relative changes in the enrichment of label in the precursor pool of glutamate and GABA itself. The first pool cannot be isolated in such experiments, so that any change in one of the glutamate pools will affect the estimation of the turnover. The possibilities of using the postmortal increase of GABA for turnover determinations have been discussed. At present there are too few data to make any conclusion on its applicability. One of the disadvantages is the short apparent lifetime (4 min) of a specific glutamate precursor pool.

There are large gaps in our knowledge on the quantitative contributions of the different synthetic and metabolic routes for glutamate. This complicates any turnover studies. Pulse labeling with glucose as the precursor and short observation time may be the most promising approach. An alternative method, which may at least indicate quantitative changes in turnover, is to limit the rate of glutamate synthesis, e.g., hypoglycemia. Under such conditions a decrease in glutamate level may indicate an increase in turnover.

There is no satisfactory model to explain the kinetic data for aspartate turnover. The level of aspartate is very sensitive to small changes in the concentrations of oxaloacetic acid.

The turnover of glycine can in principle be monitored from the incorporation of label in serine and glycine after infusion with $[3,4\text{-}^{14}C]$–glucose or after intracerebral injection of labeled serine. There are no reports in the literature on a serious evaluation of this approach after pharmacological treatment of the glycine system.

References

Altschuler R. A., Mosinger J. L., Harmison G. G., Parakkal M. H., and Wenthold R.J. (1982) Aspartate aminotransferase-like immunoreactivity as a marker for aspartate/glutamate in guinea pig photoreceptors. *Nature* (Lond.) **298**, 657–659.

Altschuler F. R., Neises G. R., Harmison G. G., Wenthold R. J., and Fex J. (1981) Immunocytochemical localization of aspartate aminotransferase immunoreactivity in cochlear nucleus of the guinea-pig. *Proc. Natl. Acad. Sci. USA* **78**, 6553–6557.

Aprison M. H. and Nadi N. S. (1978) Glycine: Inhibition from the Sacrum to the Medulla, in *Amino Acids as Chemical Transmitters* (Fonnum, F., ed.), pp. 531–571, Plenum, New York.

Balazs R., Machiyama Y., Hammond B. J., Julian T., and Richter D. (1970) The operation of the γ-aminobutyrate bypath of the tricarboxylic acid cycle in brain tissue in vitro. *Biochem. J.* **116**, 445–467.

Baxter C. F. (1976) Some Recent Advances in Studies of GABA Metabolism and Compartmentation, in *GABA in Nervous System Function* (Roberts E., Chase T.N., and Bower D.B. eds.), Raven Press, New York.

Berl S. and Clarke D. D. (1978) Metabolic Compartmentation of the Glutamate–Glutamine System; Glial Contribution, in *Amino Acids as Chemical Transmitters* (Fonnum, F., ed.) pp. 691–708, Plenum, New York.

Berl S., Clarke D. D., and Nicklas W. J. (1970) Compartmentation of citric acid cycle metabolism in brain. *J. Neurochem.* **17**, 999–1007.

Berl S., Lajhta A., and Waelsch H. (1961) Amino acid and protein metabolism. VI. Cerebral compartments of glutamic acid metabolism. *J. Neurochem.* **7**, 186–197.

Bernasconi R., Maitre L., Martin P., and Raschdorf F. (1982) The use of inhibitors of GABA-transaminase for the determination of GABA turnover in mouse brain regions: an evaluation of aminooxyacetic acid and Gabaculine. *J. Neurochem.* **38**, 57–66.

Bertilsson L., Chi-Chiang M., and Costa E. (1977) Application of principles of steady-state kinetics to the estimation of γ-aminobutyric acid turnover rate in nuclei of rat brain. *J. Pharmacol. Exp. Ther.* **200**, 277–284.

Bradford H. F. and Ward H. K. (1976) On glutaminase in mammalian synaptosomes. *Brain. Res.* **110**, 115–125.

Bradford H. F., Ward H. K., and Tomas A. (1978) Glutamine —a major substrate for nerve endings. *J. Neurochem.* **30**, 1453–1449.

Butterworth R. F., Merkel A. D., and Landreville F. (1982) Regional amino acid distribution in relation to function in insulin hypoglycemia. *J. Neurochem.* **38**, 1483–1489.

Casu M. and Gale K. (1981) Intracerebral injection of gamma vinyl–GABA: method for measuring rates of GABA synthesis in specific brain regions in vivo. *Life Sci.* **29**, 681–688.

Cattabeni F., Bugatti A., Groppetti A., Maggi A., Parenti M., and Racagni G. (1979) GABA and dopamine: Their Mutual Regulation in the Nigro-Striatal System, in *GABA Neurotransmitters* (Krogsgaard-

Larsen P., Scheel-Krüger J., and Kofod H., eds.), pp. 107–117, Munksgaard, Copenhagen.

Chapman A. G., Riley K., Evans M. C., and Meldrum B. S. (1982) Acute effects of sodium valproate and γ-vinyl–GABA on regional amino acid metabolism in the rat brain: Incorporation of 2-[^{14}C]–glucose into amino acids. *Neurochem. Res.* **7**, 1089–1105.

Cheng S.-C. (1972) Compartmentation of Tricarboxylic Acid Cycle Intermediates and Related Metabolites, in *Metabolic Compartmentation in the Brain* (Balazs R. and Cremer J.E., eds.), pp. 107–118. MacMillan, London.

Clarke D. D., London J., and Garfinkel D. (1978) Computer Modeling as an Aid to Understanding Metabolic Compartmentation of the Krebs Cycle in Brain Tissue, in *Amino Acids as Chemical Transmitters* (Fonnum F., ed.), pp. 725–738, Plenum, New York.

Collins G. G. S. (1972) GABA-2-oxoglutarate transaminase, glutamate decarboxylase, and the half-life of GABA in different areas of rat brain. *Biochem. Pharmacol.* **21**, 2849–2853.

Collins G. G. S. (1977) On the Role of Taurine in the Mammalian Central Nervous System, in *Essays in Neurochemistry and Neuropharmacology*, Vol. 1 (Youdini M.B.H., Lovenberg W., Sharman D.F., and Lagnado J.R., eds.), John Wiley, London.

Collins G. G. S. and Probett G. A. (1981) Aspartate and not glutamate is the likely transmitter of the rat lateral olfactory tract fibers. *Brain Res.* **201**, 231–234.

Costa E., Cheney D. L., Mao C. C., and Moroni F. (1978) Action of antischizophrenic drugs on the metabolism of γ-aminobutyric acid and acetylcholine in globus pallidus striatum and n. accumbens. *Fed. Proc.* **37**, 2408–2414.

Costa E., Guidotti A., Moroni F., and Peralta E. (1979). Glutamic Acid as a Transmitter Precursor and as a Transmitter, in *Advances in Biochemistry and Physiology* (Filer L.J. et al., eds.), Raven, New York.

Cremer J. E., Heath D. F., Patel A. J., Balazs R., and Cavanagh J. B. (1975) An Experimental Model of CNS Changes Associated with Chronic Liver Disease: Portocaval Anastomosis in the Rat, in *Metabolic Compartmentation and Neurotransmission* (Berl S., Clarke D.D., and Schneider D., eds.), pp. 461–478, Plenum, New York.

Cremer J. E. and Lucas H. M. (1971) Sodium pentobarbitone and metabolic compartments in rat brain. *Brain Res.* **35**, 619–621.

Cremer J. E., Sarna G. S., Teal H. M., and Cunningham V. J. (1978) Amino Acid Precursors: Their Transport into Brain and Initial Metabolism, in *Amino Acids as Chemical Transmitters* (Fonnum F., ed.), pp. 669–689, Plenum, New York.

Dann O. T. and Carter T. E. (1964) Cycloserine inhibition of gamma-amino butyric-alpha-ketoglutaric transaminase. *Biochem. Pharmacol.* **13**, 677–684.

Davidoff R. A., Graham L. T., Shank R. P., Werman R., and Aprison M. H. (1967) Changes in amino acid concentrations associated with loss of spinal interneurons. *J. Neurochem.* **14**, 1025–1031.

Engelsen B. and Fonnum F. (1983) Effects of hypoglycemia on the transmitter pool and the metabolic pool of glutamate in rat brain. *Neurosci. Lett.* **42**, 317–322.

Fahn S. (1976) Regional Distribution Studies of GABA and Other Putative Neurotransmitters and Their Enzymes, in *GABA in Nervous Systems* (Roberts, E., Chase T.N., and Tower D.B., eds.), pp. 169–186, Raven, New York.

Fitzpatrick S. M., Cooper A. J. L., and Duffy T. E. (1983) Use of β-methylene-D,L-aspartate to assess the role of aspartate aminotransferase in cerebral oxidative metabolism. *J. Neurochem.* **41**, 1370–1383.

Flethcher A. and Fowler L. (1980) γ-Aminobutyric acid metabolism in rat brain following chronic oral administration of ethanolamine O-sulfate. *Biochem. Pharmacol.* **29**, 1451–1454.

Fonnum F. (1968) The distribution of glutamate decarboxylase and aspartate transaminase in subcellular fractions of rat and guinea-pig brain. *Biochem. J.* **106**, 401–412.

Fonnum F. (ed.) (1978) *Amino Acids as Chemical Transmitters*, NATO Advanced Study Institutes Series, Series A, Life Sciences; Vol. 16. Plenum, New York.

Fonnum F. (1981) The Turnover of Transmitter Amino Acids with Special Reference to GABA, in *Central Neurotransmitter Turnover* (Pycock C.J. and Taberner P.V., eds.), pp. 105–124, University Park Press, Baltimore.

Fonnum F. (1984) Glutamate: A neurotransmitter in mammalian brain. *J. Neurochem.* **42**, 1–11.

Fonnum F., Storm-Mathisen J., and Divac I. (1981) Biochemical evidence for glutamate as neurotransmitter in the corticostriatal and cortico-thalamic fibers in rat brain. *Neuroscience.* **6**, 863–875.

Fonnum F. and Walberg F. (1973) An estimation of the concentration of γ-aminobutyric acid and glutamate decarboxylase in the inhibitory Purkinje axon terminals of the cat. *Brain Res.* **54**, 115–127.

Forchetti C. M., Marco M. J., and Meek J. L. (1982) Serotonin and γ-aminobutyric acid turnover after injection into the median raphe and Substance P and D-ala-met-enkephalin amide. *J. Neurochem.* **38**, 1383–1386.

Fowler J. and John R. A. (1972) Active-site-directed irreversible inhibition of rat brain 4-aminobutyrate aminotransferase by ethanolamine-O-sulphate in vitro and in vivo. *Biochem. J.* **130**, 569–573.

Freeman M. E., Lane J. D., and Smith J. E. (1983) Turnover rate of amino acid neurotransmitters in regions of rat cerebellum. *J. Neurochem.* **40**, 1441–1447.

Fuxe K., Andersson K., Ogren S. O., Perez de la Mora M., Schwartz R., Hökfelt T., Eneroth P., Gustafsson J.-Å, and Skeff P. (1979) GABA Neurons and Their Interaction with Monoamine Neurons. An Anatomical, Pharmacological, and Functional Analysis, in *GABA Neurotransmitters* (Krogsgaard-Larsen P., Scheel-Krüger J., and Kofod H., eds.), pp. 74–94, Munksgaard, Copenhagen.

Gaitonde M. K., Evison E., and Evans G. M. (1983) The rate of utilization of glucose via hexosemonophosphate shunt in brain. *J. Neurochem.* **41**, 1253–1260.

Gauchy M. L., Kemel M. L., Glowinski J., and Besson M. J. (1980) In vivo release of endogenously synthesized [^{3}H]–GABA from the cat substantia nigra and the pallidoentopeduncular nuclei. *Brain Res.* **193**, 129–142.

Geddes J. W. and Wood J. D. (1984) Changes in the amino acid content of nerve endings (synaptosomes) induced by drugs that alter the metabolism of glutamate and γ-aminobutyric acid. *J. Neurochem.* **42**, 16–24.

Giorgi O. and Meek J. L. (1984) γ-aminobutyric acid turnover in rat striatum: Effects of glutamate and kainic acid. *J. Neurochem.* **42**, 215–220.

Graham L. T. Jr. and Aprison M. H. (1969) Distribution of some enzymes associated with the metabolism of aspartate, glutamate, γ-aminobutyrate, and glutamine in cat spinal cord. *J. Neurochem.* **16**, 559–566.

Guiener R., Markovitz D., Huxtable R., and Bressler R. (1975) Excitability modulation by taurine. Transmembrane measurements of neuromuscular transmission. *J. Neurol. Sci.* **24**, 351–359.

Hamberger A., Chiang G. H., Nylén E. S., Scheff S. W., and Cotman C. W. (1979a) Glutamate as a CNS transmitter. I. Evaluation of glucose and glutamine as precursors for the synthesis of preferentially released glutamate. *Brain Res.* **168**, 513–530.

Hamberger A., Chiang G. H., Sandoval E., and Cotman C. W. (1979b) Glutamate as a CNS transmitter. II. Regulation of synthesis in the releasable pool. *Brain Res.* **168**, 531–541.

Hawkins R. A., Mans A. M., Davies D. W., Hubbard L. S., and Lu D. M. (1983) Glucose availability to individual cerebral structures is correlated to glucose metabolism. *J. Neurochem.* **40**, 1013–1018.

Hökfelt T., Ljungdahl A., Perez do la Mora M., and Fuxe K. (1976) Further evidence that apomorphine increases GABA turnover in the DA-cell-body rich and DA-nerve-terminals rich areas of the brain. *Neurosci. Lett.* **2**, 239–242.

Ionesco M. J. and Gale K. (1979) Dissociation between drug-induced increases in nerve terminal and non-nerve terminal pools of GABA in vivo. *Eur. J. Pharmacol.* **59**, 125–129.

Johnson J. L. (1978) The excitant amino acids, glutamic, and aspartic acids, as transmitter candidates in the vertebrate central nervous system. *Prog. Neurobiol.* **10**, 155–202.

Jung M. J., Lippert B., Metcalf B. W., Böhlen P., and Schlechter P. J. (1977a) γ-Vinyl-GABA (4-aminohexionic acid), a new selective irreversible inhibitor of GABA-T: Effects on brain GABA metabolism in mice. *J. Neurochem.* **29**, 797–802.

Jung M. J., Lippert B., Metcalf B. W., Schechter P. J., Böhlen P., and Sjoerdsma A. (1977b) The effect of 4-aminohex-5-ionic acid (γ-acetylenic GABA, γ-ethynyl GABA), a catalytic inhibitor of GABA transaminase, on brain GABA metabolism in vivo. *J. Neurochem.* **28**, 717–723.

Karlsson G., Fonnum F., Malthe-Sørenssen D., and Storm-Mathisen J. (1974) Effect of the convulsive agent 3-mercaptopropionic acid on the levels of GABA, other amino acids, and glutamate decarboxylase in different regions of rat brain. *Biochem. Pharmacol.* **23**, 3053–3061.

Kataoka K., Bak I. J., Hassler R., Kim J. S., and Wagner A. (1974) L-glutamate decarboxylase and cholineacetyltransferase activity in the substantia nigra and the striatum after surgical interruption of the strio-nigral fibers of the baboon. *Exp. Brain Res.* **19**, 217–227.

Knierem K. M., Medina M. A., and Stavinoha W. B. (1977) The levels of GABA in mouse brain following tissue inactivation by microwave irradiation. *J. Neurochem.* **28**, 885–886.

Kontro P., Marnela K. M., and Oja S. S. (1980) Free amino acids in the synaptosome and synaptic vesicle fractions of different bovine brain areas. *Brain Res.* **184**, 129–141.

Korf J. and Venema K. (1983) Amino acids in the substantia nigra in rats with striatal lesions produced by kainic acid. *J. Neurochem.* **40**, 1171–1173.

Kvamme E. and Olsen B. E. (1980) Substrate mediated regulation of phosphate-activated glutaminase in nervous tissue. *Brain Res.* **181**, 228–233.

Lane J. D., Sands M. P., Freeman M. E., Cherek D. R., and Smith J. E. (1982) Amino acid neurotransmitter utilization in discrete rat brain regions is correlated with conditioned emotional response. *Pharmacol. Biochem. Behav.* **16**, 329–340.

Leach M. J., Miller A. A., O'Donell R. A., and Webster R. A. (1983) Reduced cortical glutamine concentrations in electrically kindled rats. *J. Neurochem.* **41**, 1492–1494.

Leong S. F., Lai J. C. K., Lim L., and Clark J. B. (1984) The activities of some energy-metabolizing enzymes in nonsynaptic (free) and synaptic mitochondria derived from selected brain regions. *J. Neurochem.* **42**, 1306–1312.

Löscher W. (1980) Effect of inhibitors of GABA transaminase on the synthesis, binding, uptake, and metabolism of GABA. *J. Neurochem.* **34**, 1603–1608.

Löscher W. and Vetter M. (1984) Relationship between drug-induced increases of GABA levels in discrete brain areas and different pharmacological effects in rats. *Biochem. Pharmacol.* **33**, 1907–1914.

Lund-Karlsen R. and Fonnum F. (1978) Evidence for glutamate as a neurotransmitter in the corticofugal fibers to the dorsal lateral geniculate body and the superior colliculus in rats. *Brain Res.* **151**, 457–467.

Machiyama Y., Balazs R., Hammond B. J., Julian T., and Richer D. (1970) The metabolism of GABA and glucose in potassium stimulated brain tissue in vitro. *Biochem J.* **116,** 469–482.

Mangan S. L. and Whittaker V. P. (1966) The distribution of free amino acids in subcellular fractions of guinea pig brain. *Biochem. J.* **98,** 128–137.

Mansky T., Mestres-Ventura P., and Wuttke W. (1981) Involvement of GABA of the feedback action of estradiol on gonadrotropin and prolactin release: hypothalamic GABA and catecholamine turnover rates. *Brain Res.* **231,** 353–364.

Mansky T., Düker E., and Wuttke W. (1983) Hypothalamic and limbic GABA concentrations and turnover rates and glutamate concentrations following induction of hyperprolactinemia in ovariectomized rats. *Neurosci. Lett.* **39,** 167–272.

Mao C. C., Peralta E., Morini F., and Costa E. (1978) The turnover rate of γ-aminobutyric acid in the substantia nigra following electrical stimulation or lesioning of the striatonigral pathways. *Brain Res.* **155,** 147–152.

Marco E., Mao C. C., Revuelta A., Peralta E., and Costa E. (1978) Turnover rates of γ-aminobutyric acid in substantia nigra, n. caudatus, globus pallidus, and n. accumbens of rat injected with cataleptogenic and noncataleptogenic antipsychotics. *Neuropharmacol.* **17,** 589–596.

Martinez-Hernandez A., Bell K. P., and Norenberg M. D. (1977) Glutamine synthetase: Glial localization in brain. *Science* **195,** 1356–1358.

Massieu G. H., Tapia R., Pasantes H., and Orteg B. G. (1964) Convulsant effect of L-glutamic acid γ-hydrazine by simultaneous treatment with pyridoxal phosphate. *Biochem. Pharmacol.* **13,** 118–120.

Matsui Y. and Deguchi T. (1977) Effects of Gabaculine, a new potent inhibitor of gamma-aminobutyrate transaminase, on the brain gamma-aminobutyrate content and convulsions in mice. *Life Sci.* **20,** 1291–1296.

McGeer E. G. and McGeer P. L. (1978) Localization of glutaminase in the rat neostriatum. *J. Neurochem.* **32,** 1071.

McLaughlin B. J., Wood J., Saito K., Barber R., Vaughn J. E., Roberts E., and Wu J.-Y. (1974) The fine structural localization of glutamate decarboxylase in synaptic terminals of rodent cerebellum. *Brain Res.* **76,** 377–391.

Melis M. R. and Gale K. (1983) Effect of dopamine agonists on γ-aminobutyric acid (GABA) turnover in the superior colliculus: Evidence that nigrotectal GABA projections are under influence of dopaminergic transmission. *J. Pharmacol. Exp. Ther.* **226,** 425–431.

Metcalf B. W. (1979) Inhibitors of GABA metabolism. *Biochem. Pharmacol.* **28,** 1705–1712.

Miller L. P., Martin D. L., Mazumder A., and Watkins J. R. (1978) Studies on the regulation of GABA synthesis: Substrate-promoted dissociation of pyridoxal-P from GAD. *J. Neurochem.* **30,** 361–369.

Minard F. N. and Mushahwar I. K. (1966) Synthesis of γ-aminobutyric acid from a pool of glutamic acid in brain after decapitation. *Life Sci.* **5,** 1409–1413.

Minchin M. C. W. and Beart P. M. (1974) Compartmentation of amino acid metabolism in the rat dorsal ganglia: A metabolic and autoradiographic study. *Brain Res.* **83,** 437–449.

Minchin M. C. W. and Fonnum F. (1979) The metabolism of GABA and other amino acids in rat substantia nigra slices following lesions of the striatonigral pathway. *J. Neurochem.* **32,** 203–210.

Moroni F., Cheney D. L., Peralta E., and Costa E. (1978) Opiate receptor agonists as modulators of γ-aminobutyric acid turnover in the nucleus caudatus, globus pallidus, and substantia nigra. *J. Pharmacol. Exp. Ther.* **207,** 870–877.

Moroni F., Lombardi G., Moneti G., and Cortesini C. (1983) The release and neosynthesis of glutamic acid are increased in experimental models of hepatic encephalopathy. *J. Neurochem.* **40,** 850–854.

Nadler J. V. and Smith E. M. (1981) Perforant path lesion depletes glutamate content of fascia dentata synaptosomes. *Neurosci. Lett.* **3,** 275–280.

Nadler J. V., White W. F., Vaca K. W., Perry B. W., and Cotman C. W. (1978) Biochemical correlates of transmission mediated by glutamate and aspartate. *J. Neurochem.* **31,** 147–155.

Neff N. H., Spano P. F., Gropetti A., Wang C. T., and Costa E. (1971) A simple procedure for calculating the synthesis rate of norepinephrine, dopamine, and serotonin in rat brain. *J. Pharmacol. Exp. Ther.* **176,** 701–710.

Nicklas W. J. (1983) Relative Contributions of Neurons and Glia to Metabolism of Glutamate and GABA, in *Glutamine, Glutamate, and GABA in the Central Nervous System* (Hertz L., Kvamme E., McGeer E. G., and Schousboe A., eds.), pp. 219–231, Alan R. Liss, New York.

Nicklas W. J., Nunez R., Berl S., and Duvoisin R. (1979) Neuronal glial contributions to transmitter amino acid metabolism: Studies with kainic acid-induced lesions of rat striatum. *J. Neurochem.* **33,** 839–844.

Norenberg M. D. and Martinez-Hernandez A. (1979) Fine structural localization of glutamine synthetase in astrocytes of rat brain. *Brain Res.* **161,** 303–310.

Okada Y. (1982) Fine localization of GABA (γ-Aminobutyric Acid) and GAD (Glutamate Decarboxylase) in a Single Deiters Neuron — Significance of the Uneven Distribution of GABA and GAD in the CNS, in *Problems in GABA Research from Brain to Bacteria* (Okada Y. and Roberts E., eds.), pp. 30–54, Excerpta Medica, Amsterdam.

Okada Y. and Roberts E. (1982) *Problems in GABA Research from Brain to Bacteria.* Excerpta Medica, Amsterdam.

Omholt-Jensen B. (1984) Metode for måling av TR-GABA ved hemming av GABA-T med GVG. Thesis Oslo University, Dept. of Pharmacy.

Pagliusi S. R., Gomes C., Leite J. R., and Trolin G. (1983) Amino-oxyacetic acid-induced accumulation of GABA in the rat brain. Interaction with GABA receptors and distribution in compartments. *Naunyn-Schmiedeberg's Arch. Pharmacol.* **322,** 210–215.

Palfreyman M. G., Schechter P. J., Buckett W. R., Tell G. P., and Koch Weser J. (1981) The pharmacology of GABA-transaminase inhibitors. *Biochem. Pharmacol.* **30,** 817–824.

Patel J. A. (1982) The Distribution and Regulation in Nerve Cells and Astrocytes of Certain Enzymes Associated with the Metabolic Compartmentation of Glutamate, in *Neurotransmitter Interaction and Compartmentation* (Bradford A.F., ed.), pp. 411–429, Plenum, New York.

Patel A. J., Balazs R., and Richter D. (1970) Contribution of the GABA bypath to glucose oxidation, and the development of compartmentation in the brain. *Nature* (Lond.) **226,** 1160–1161.

Patel A. J., Johnsen A. L., and Balazs R. (1974) Metabolic compartmentation of glutamate associated with the formation of γ-aminobutyrate. *J. Neurochem.* **23,** 1271–1279.

Perez de la Mora M., Fuxe K., Hökfelt T., and Ljungdahl Å. (1977) Evidence for an impulse-dependent GABA accumulation in the substantia nigra after treatment with γ-glutamyl-hydrazide. *Neurosci. Lett.* **5,** 75–82.

Pericic D., Eng N., and Walters J. R. (1978) Postmortem and aminooxyacetic acid-induced accumulation of GABA: Effect of γ-butyrolactone and picrotoxin. *J. Neurochem.* **30,** 767–774.

Porter T. G. and Martin O. L. (1984) Evidence for feedback regulation of glutamate decarboxylase by γ-aminobutyric acid. *J. Neurochem.* **43,** 1464–1467.

Racagni G., Cheney D. L., Trabucchi M., Wang C. T., and Costa E. (1974) Measurements of acetylcholine turnover rate in discrete areas of rat brain. *Life Sci.* **15,** 1961–1975.

Raff M. C., Miller R. H., and Noble M. (1983) A glial progenitor cell that develops in vitro into a astrocyte or an oligodendrocyte depending on culture medium. *Nature* (Lond.) **303,** 390–392.

Rando R. R. and Bangerter F. W. (1977) The in vivo inhibition of GABA transaminase by Gabaculine. *Biochem. Biophys. Res. Commun.* **76,** 1276–1281.

Rassin D. L. (1972) Amino acids as putative transmitters: Failure to bind to synaptic vesicles of guinea-pig cerebral cortex. *J. Neurochem.* **19,** 139–148.

Rassin D. K., Sturman J. A., and Gaull G. E. (1981) Sulfur amino acid metabolism in the developing rhesus monkey brain: Subcellular studies of taurine, cysteinesulfinic acid decarboxylase, γ-amino-butyric acid, and glutamic acid decarboxylase. *J. Neurochem.* **37,** 740–748.

Reijnierse G. L. A., Veldstra H., and van den Berg C. J. (1975) Subcellular localization of γ-aminobutyrate transaminase and glutamate dehydrogenase in adult rat brain. *Biochem. J.* **152,** 469–475.

Reubi J. C. (1980) Comparative study of the release of glutamate and GABA, newly synthesized from glutamine in various regions of the central nervous system. *Neurosci.* **5,** 2145–2150.

Revuelta A. V., Cheney D. L., and Costa E. (1981) Measurements of γ-Aminobutyric Acid Turnover Rates in Brain Nuclei as an Index of Interactions Between γ-Aminobutyric Acid and Other Transmitters, in *Glutamate as a Neurotransmitter* (Di Chiara G. and Gessa G. L., eds.), Raven, New York.

Salganicoff L. and De Robertis E. (1965) Subcellular distribution of the enzymes of the glutamic acid, glutamine, and γ-aminobutyric acid cycles in brain. *J. Neurochem.* **12,** 287–309.

Sandberg M., Ward H. K., and Bradford H. F. (1985) Effect of corticostriate pathways lesion on the activities of enzymes involved in synthesis and metabolism of amino acid neurotransmitters in the striatum. *J. Neurochem.* **44,** 42–47.

Sarhan S. and Seiler N. (1979) Metabolic inhibitions and subcellular distribution of GABA. *J. Neurosci. Res.* **4,** 399–421.

Schon F. and Kelly J. S. (1974) Autoradiographic localization of [^{3}H]-GABA and [^{3}H]-glutamate over satellite glial cells. *Brain Res.* **66,** 275–288.

Schousboe A., Svenneby G., and Hertz L. (1977) Uptake and metabolism of glutamate in astrocytes cultured from dissociated mouse brain hemispheres. *J. Neurochem.* **29,** 299.

Shank R. P. and Aprison M. H. (1970) The metabolism in vivo of glycine and serine in eight areas of the rat central nervous system. *J. Neurochem.* **17,** 1461–1478.

Shank R. P., Aprison M. H., and Baxter M. H. (1973) Precursors of glycine in the central nervous system: Comparison of specific activities in glycine and other amino acids after administration of (U-^{14}C)–glucose, (3,4-^{14}C)–glucose, (1-^{14}C)–glucose, (U-^{14}C)–serine, or (1,5-^{14}C)–citrate to the rat. *Brain Res.* **51,** 301–308.

Shank R. P. and Campbell G. LeM. (1982) Glutamine and alpha-ketoglutarate uptake and metabolism by nerve terminal enriched material from mouse cerebellum. *Neurochem. Res.* **7,** 601–616.

Shank R. P. and Campbell G. LeM. (1984) α-Ketoglutarate and malate uptake and metabolism by synaptosomes: Further evidence for an astrocyte-to-neuron metabolic shuttle. *J. Neurochem.* **42,** 1153–1161.

Shank R. P., Campbell G. LeM., Freitag S. U., and Utter M. F. (1981) Evidence that pyruvate carboxylase is an astrocyte-specific enzyme. *Soc. Neurosci. Abstr.* **7,** 936.

Smith J. E., Co C., Freeman M. E., and Lane J. D. (1982) Brain neurotransmitter turnover correlated with morphine-seeking behavior of rats. *Pharmacol. Biochem. Behav.* **16,** 509–519.

Sterri S. H. and Fonnum F. (1980) Acetyl-CoA-synthesizing enzymes in cholinergic nerve terminals. *J. Neurochem.* **35,** 249–254.

Thanki C. M., Sugden D., Thomas N. J., and Bradford H. F. (1983) In vivo release from cerebral cortex of [^{14}C]-glutamate synthesized from [U-^{14}C]-glutamate. *J. Neurochem.* **41,** 611–617.

van den Berg C. J. (1973) A Model of Compartmentation in Mouse Brain Based on Glucose and Acetate Metabolism, in *Metabolic Compartmentation in the Brain* (Balazs R. and Cremer J. E., eds.), pp. 137–166, MacMillan,

van den Berg C. J. and Garfinkel D. (1971) A simulation study of brain compartments. *Biochem. J.* **123,** 211–218.

van den Berg C. J., Matheson D. F., and Nijenmanting W. C. (1978) Compartmentation of Amino Acids in Brain: The GABA-Glutamine-Glutamate Cycle, in *Amino Acids as Neurotransmitters* (Fonnum, F., ed.), pp. 709–723, NATO Advanced Study Institutes Series Series A, *Life Sciences,* Vol. 16, Plenum, New York.

van den Berg C. J., Matheson D. F., Ronda G., Reijnierse G. L. A., Blokhuis G. G. D., Kroon M. C., Clarke D. D., and Garfinkel D. (1975) A Model of Glutamate Metabolism in Brain: A Biochemical Analysis of a Heterogenous Structure, in *Metabolic Compartmentation and Neurotransmission,* (Berl S., Clarke D.D., and Schneider D., eds.), pp. 709–723, NATO Advanced Study Institute Series, Vol. 16, Plenum, New York.

van der Heyden J. A. M., de Kloet E. R., Korf J., and Versteeg D. H. G. (1979) GABA content of discrete brain nuclei and spinal cord of the rat. *J. Neurochem.* **33,** 857–861.

van der Heyden J. A. M. and Korf J. (1978) Regional levels of GABA in the brain: Rapid semiautomated assay and prevention of postmortem increase by 3-mercaptopropionic acid. *J. Neurochem.* **31,** 197–203.

van Gelder N. M. (1966) The effect of aminooxyacetic acid on the metabolism of γ-aminobutyric acid in brain. *Biochem. Pharmacol.* **15,** 533–539.

van Kempen G. M. J., van den Berg C. J., van der Helm H. J., and Veldstra H. (1965) Intracellular localization of glutamate decarboxylase, γ-aminobutyrate transaminase, and some other enzymes in brain tissue. *J. Neurochem.* **12,** 581–588.

Vincent S. R., Kimura H., and McGeer E. G. (1980) The pharmacohistochemical demonstration of GABA transaminase. *Neurosci. Lett.* **8,** 354–358.

Walaas I. and Fonnum F. (1980) Biochemical evidence for glutamate as a transmitter in hippocampal efferents to the basal forebrain and hypothalamus in rat brain. *Neuroscience* **5,** 1691–1698.

Wallach D. P. (1961) Studies on the GABA pathway. I. The inhibition of γ-aminobutyric acid-α-ketoglutaric acid transaminase in vitro and in vivo by U-7524 (aminooxyacetic acid). *Biochem. Pharmacol.* **5,** 323–331.

Walters J. R., Eng N., Pericic D., and Miller L. P. (1978) Effects of aminooxyacetic acid and L-glutamic acid-γ-hydrazide on GABA metabolism in specific brain regions. *J. Neurochem.* **30,** 759–766.

Ward H. W., Thanki C. M., and Bradford H. F. (1983) Glutamine and glucose as precursors of transmitter amino acids: ex vivo studies. *J. Neurochem.* **40,** 855–860.

Ward H. W., Thanki C. M., Peterson D. W., and Bradford H. F. (1982) Brain glutaminase activity in relation to transmitter glutamate biosynthesis. *Biochem. Soc. Trans.* **10,** 369–370.

Wenthold R. J. (1979) Release of endogenous glutamic acid, aspartic acid, and GABA from cochlear nucleus slices. *Brain Res.* **162,** 338–343.

Wenthold R. J. and Altschuler R. A. (1983) Immunocytochemistry of Aspartate Aminotransferase and Glutaminase, in *Glutamine, Glutamate, and GABA in the Central Nervous System* (Hertz L., Kvamme E., McGeer E., and Schousboe A., eds.), pp. 33–50, Alan Liss, New York.

Westerberg E., Chapman A. G., and Meldrum B. S. (1983) Effect of 2-amino-7-phosphonoheptanoic acid on regional brain amino acid levels in fed and fasted rodents. *J. Neurochem.* **41,** 1755–1760.

Wiklund L., Toggenburger G., and Cuenod M. (1982) Aspartate: Possible neurotransmitter in cerebellar climbing fibers. *Science* **216,** 78–80.

Wood J. D. (1981) Evaluation of a synaptosomal model for monitoring in vivo changes in the GABA and glutamate content of nerve endings. *Int. J. Biochem.* **13,** 543–548.

Wood J. D. and Kurylo E. (1984) Amino acid content of nerve endings (synaptosomes) in different regions of brain: Effects of gabaculine and isonicotinic acid hydrazide. *J. Neurochem.* **42,** 420–425.

Wood J. D., Russell M. P., and Kurylo E. (1980) The γ-aminobutyrate content of nerve endings (synaptosomes) in mice after the instramuscular injection of γ-aminobutyrate-elevating agents: A possible role in anticonvulsant activity. *J. Neurochem.* **35,** 125–130.

Wu J.-Y. (1982) Purification and characterization of cysteic/cysteine sulphinic acids decarboxylase and L-glutamate decarboxylase in bovine brain. *Proc. Natl. Acad. Sci. USA* **79,** 4270–4274.

Yamatsu K., Yamanishi Y., Ikeda M., Uzuo T., and Okada Y. (1982) Postmortem GABA Increase in Discrete Regions of the Rat Brain —Involvement of GAD and GABA-T Activity, in *Problems in GABA from Brain to Bacteria* (Okada Y. and Roberts E., eds.), pp. 30–40, Excerpta Medica, Amsterdam.

Yoneda Y., Roberts E., and Dietz G. W., Jr. (1982) A new synaptosomal biosynthetic pathway of glutamate and GABA from ornithine and its negative feedback inhibition by GABA. *J. Neurochem.* **38,** 1686–1694.

Yu A. C. H., Drejer J., Hertz L., and Schousboe A. (1983) Pyruvate carboxylase activity in primary cultures of astrocytes and neurons. *J. Neurochem.* **41,** 1484–1487.

Chapter 10

Uptake and Release of Amino Acid Neurotransmitters

WOLFGANG WALZ

1. Introduction

The amino acid content of the central nervous system (CNS) is controlled by the blood–brain barrier. Despite a constant exchange of amino acids in both directions, this interface mediates a net uptake from blood plasma into cerebrospinal fluid (CSF). As in most other cells, amino acids in the brain have their roles as constituents of protein, metabolic precursors, and intermediates in energy and nitrogen metabolism; furthermore these compounds are involved in osmoregulation. The system in which these functions are most fully understood is probably the Ehrlich ascites tumor cell (Johnstone, 1979). In the brain, some amino acids clearly have an additional function; that of neurotransmitter, i.e., a chemical messenger that bridges the synaptic clefts between neural membranes, thus mediating interneuronal signaling. Glutamate and aspartate are established excitatory transmitters, whereas this role for cysteate and cysteine sulfinate is putative. Inhibitory transmitters are γ-aminobutyric acid (GABA), glycine (in the spinal cord), and taurine (regarded as a more general neuromodulator). Glutamine is not a neurotransmitter, but an important transport vehicle to transfer neurotransmitters in an inactive form between cells. This compound is, therefore, also to be considered in the present context.

This review focuses on experimental paradigms for investigating the involvement of amino acids in CNS neurotransmission, i.e., their release in response to electrical activation of the presynaptic membrane and their inactivation and/or removal from the extracellular space (ECS) to terminate their signal function. An introduction to amino acid transport for the general function of nerve cells not related to signal transmission may be found elsewhere (Oja and Korpi, 1983). Autoradiography is a major methodology in the investigation of amino acid uptake and is dealt with in a separate chapter in this volume (Scott-Young, 1985). The analysis and identification of amino acids in various samples is also discussed in several chapters of this volume. It should be noted that in recent years some excellent reviews of amino acid function as neurotransmitters have been published (Hertz, 1979; Hertz et al., 1983; Schousboe, 1981, 1982; Lajtha, 1983). Some of these have emphasized methods, and the interested reader should consult these sources for additional information and alternative views (Fagg and Lane, 1979; Orrego, 1979; Cox and Bradford, 1978; Bradford, 1981).

2. Problems in the Transport of Amino Acids During Neurotransmission

There are two main questions concerning transport of amino acids during neurotransmission: Is an amino acid used by a particular pathway as the neurotransmitter and, if so, what are the mechanisms of release, action, and reuptake?

Classically the release of a transmitter is by nerve terminals via stimulus–secretion coupling, which is calcium-dependent (Rubin, 1970). Therefore, studies have been conducted in an attempt to demonstrate that a group of neurons may release an amino acid in response to a depolarizing electrical stimulus in a calcium-dependent way. This is methodologically very difficult. One reason is the heterogeneity of the preparation. For example, there are reports that glial cells may also release amino acids in a calcium-dependent way (Minchin and Iversen, 1974; Roberts, 1974). Direct electrical stimulation of target neurons is difficult, and the indirect methods of depolarizing neurons that are used introduce sources of error. Spontaneous release of amino acids has also been observed; some of this release may represent homoexchange. Amino acids are not concentrated in synaptic vesicles. However, their cytoplasmic concentration is so high that exocytosis of synaptic vesicles could, theoretically, release enough amino acid molecules to account for observed postsynaptic effects. Nevertheless, the issue of whether amino acid neurotransmitters are released by exocytosis of synaptic vesicles must stay completely open, partly because synaptosomal preparations are still lacking the necessary degree of purity.

Transmitter action of amino acids is terminated by uptake into neurons and glial cells. For all amino acid neurotransmitter candidates there are, in addition to the normal low-affinity uptake found for all amino acids, high-affinity uptake systems ($K_m < 50$ μM). These uptake systems behave in an energy- and sodium-dependent manner, which is a probable indication that the inward driving force for the sodium ions is harnessed to accumulate amino acids against a gradient. The existence of a high-affinity uptake system is seen as a classical criterion for transmitter termination and, therefore, a necessary condition for the establishment of a substance as a neurotransmitter. In tissue that has no nerve terminals, however, high-affinity uptake systems have been found for nonneurotransmitter amino acids and for putative neurotransmitter amino acids. This criterion has therefore lost its original strength.

Other problems concern the role of glial cells in the termination of the transmitter action: Is their role more than that of a safety valve? Are there differences toward specific amino acids? What do glial cells do with amino acid transmitters after their uptake of them? Some researchers favor the possibility that the amino acid neurotransmitters are transferred back to neurons via the electrically inactive amino acid glutamine (Henn, 1982). Other authors favor a close neuronal and astrocytic metabolic relationship that operates in such a way that nerve endings are not able to complete the tricarboxylic acid cycle (formation of succinate, fumarate, maleate, and oxaloacetate). Glutamate, aspartate, and GABA are thus the main products of the nerve endings and are "transferred" to astrocytes that metabolize these amino acids to glutamine. This amino acid is released and transported back into nerve endings, where it serves as a metabolic precursor (Hertz, 1979). Uptake and release studies with amino acids in different systems have been used and continue to be used to resolve these questions. The important experimental designs used in these studies are described and criticized below.

3. Uptake

To study uptake of amino acids, the radiolabeled exogenous amino acid is usually incubated along with the preparation for a rectilinear period of the uptake. The preparation is then washed to remove the extracellular radioactivity and the radioactivity in the preparation is counted.

3.1. Selected Methodological Problems

Some major methodological problems are present in all uptake systems. They are described in this section.

3.1.1. Metabolism of Amino Acids Under Uptake Assay Conditions

GABA and glutamate are metabolized extensively. Especially during assays involving studies of high-affinity uptake, when low substrate concentrations are used, this factor can play a role and lead to an underestimation of the true uptake, since radioactivity may be lost from the tissue. This process depends on the kind of isotope and tissue used. The radioactivity of [14C]-GABA is largely incorporated into CO_2, aspartate, glutamate, and glutamine in subcellular fractions (Varon et al., 1975; Martin and Smith, 1972). In brain slices, substantial amounts of the radioactivity are released as [14C]-glutamine into the medium. [3H]-GABA metabolites are rapidly lost from brain slices or synaptosomes into the medium (Iversen and Neal, 1968; Levi and Raiteri, 1973) because metabolism via succinate and the tricarboxylic acid cycle results in conversion to tritiated water (Martin, 1976). In addition, metabolic degradation of transported [3H]-GABA plays a greater role in certain preparations, such as retina and isolated ganglia (Neal and Starr, 1973; Schon and Kelly, 1974; Beart et al., 1974). [14C]-Labeled metabolites derived from [14C]-glutamate are more concentrated in the incubation medium than in the tissue and most of this radioactivity is probably aspartate (Levi and Raiteri, 1973). These authors found that the experimental conditions of the incubation in which the metabolism of both GABA and glutamate can be kept sufficiently low are very restricted. They suggested that at 37°C and with an amino acid concentration of 20 μM, not more than 0.4 mg tissue protein/mL of incubation mixture should be used, and the incubation time should not exceed 10 min. Martin and Smith (1972) suggested 0.2 mg tissue protein/mL. These limits should be decreased when working with lower amino acid concentrations.

Amino-oxyacetic acid (AOAA) is an inhibitor of GABA-2-oxoglutarate aminotransferase (GABA-T) and is used successfully to prevent GABA degradation during uptake studies (Szerb, 1983). It is present during incubation with labeled GABA and while measuring release. Martin (1976) has pointed out that compounds that influence GABA metabolism may also influence the apparent uptake of GABA without directly affecting the transport system. One example is AOAA, which leads to an apparent stimulation of [3H]-GABA uptake in several preparations.

3.1.2. Loss of Amino Acids by Cold-Shock Treatment

After a preincubation time with radioactive labeled amino acids, the uptake period has to be stopped, and externally present labeled amino acids have to be removed. With synaptosomes, this is normally done by rapid cooling to 0–4°C, collection of the synaptosomes on cellulose ester filters, and washing with ice-cold medium. It was reported (Levi et al., 1976) that this cold-shock treatment would lead to loss of about 70% of the accumulated radioactivity from the incubation with labeled amino acids [GABA, glutamate, glycine, taurine, α-aminoiso-butyric acid (AIB), phenylalanine, leucine]. The loss of radioactive noradrenaline amounts to only about 20%. The authors suggested the sudden reduction in temperature causes a sudden increase in the permeability of the synaptosomal membrane for the accumulated amino acids. However, this could not be verified by other authors. Simon et al. (1974) used two different procedures to terminate the uptake period and to wash the synaptosomes. They diluted a 200 μL sample of incubated synaptosomes into 2 mL of ice-cold incubation medium, filtered the sample through a cellulose triacetate membrane filter (0.2 μm pore diameter), and rinsed the filter with 0.8 mL of incubation medium. In an alternative method, the uptake assay was undertaken in prewarmed 15 mL polypropylene centrifuge tubes and, to terminate the uptake period, the tubes were quickly placed in a warm (room temperature) Sorvall RC-2B centrifuge and accelerated to 28,000 *g*. This required about 40 s. As soon as this point was reached, the centrifuge was turned off and the rotor allowed to stop. The total time for the centrifugation was about 3 min. The supernatant was decanted and any remaining droplets of incubation medium removed with a cotton swab. The authors found no detectable difference in measured values of the uptake of [³H]-GABA between the samples rinsed with cold fluid (0°C) and warm fluid (27°C). Levi et al. (1976) reported that the effect of the cold-shock on amino acid loss from synaptosomes varied with the experimental conditions in which the shock was performed (temperature, filtration rate, and so on) and this may explain the discrepant findings.

3.1.3. Is Homoexchange of Amino Acids an In Vitro Artifact?

Several authors concluded that the high-affinity uptake system might represent a homoexchange process (Levi and Raiteri, 1974; Simon et al., 1974; Raiteri et al., 1975; Levi et al., 1976). Synaptosome preparations preloaded with [³H]-GABA or [³H]-glycine released them in a dose-dependent manner when the concentration of unlabeled GABA in the medium was increased. The release was sensitive to sodium removal and showed temperature sensitivity, and other kinetic characteristics were similar to the high-affinity uptake. Such an exchange was also demonstrated for rat sensory ganglia (Minchin, 1975; Roberts, 1976). However, Ryan and Roskoski (1977) showed that the uptake of radioactive GABA was independent of the intrasynaptosomal concentration of the amino acid. Similar results against the existence of a homoexchange component were obtained from the retina with GABA and glutamate (Lake and Voaden, 1976; White and Neal, 1976) and for glycine uptake in synaptosomes (Aprison and McBride, 1973). It is not clear what accounts for this discrepancy. Lake and Voaden (1976) suggested that the homoexchange is favored when the internal sodium concentration is raised. Roskoski (1978) found that the synaptosomal protein concentration within the medium is critical and that homoexchange occurs when it is lower than 0.5 mg/mL. Sellstrom et al. (1976) proposed that the membrane potential in isolated synaptosomes is considerably less than that of synaptic endings in vivo. Assuming that the energy for amino acid transport resides entirely in the ionic gradient, they suggest that high-affinity net uptake systems are functionally operative in intact tissue. Pastuszko et al. (1981) found only an insignificant amount of homoexchange for GABA with synaptosomes. They assumed that the isolation of synaptosomes in Ficoll density gradients yields preparations of better quality—a conclusion that is subject to criticism (Tapia, 1983). Convincing evidence that *in situ* net uptake of glutamate via a high-affinity system takes place was found by Shank and Baxter (1975) using brains from toads. Thus there is a certain degree of agreement that the homoexchange might be an in vitro phenomenon (*see*, however, Levi et al., 1978) and caution has to be taken when the uptake is measured only by the accumulation of a radiolabeled amino acid (Fagg and Lane, 1979).

3.2. Calculation of Transport Kinetic Constants

A crucial point in evaluating uptake data for amino acids is whether this uptake is mediated by a low-affinity system alone, or by a low- and a high-affinity system, i.e., if the uptake is mediated by one or two transport carriers. Shank and Campbell (1984) calculated the K_m and V_{max} values with the Pennzyme computer program (Kohn et al., 1979). This program uses a weighted nonlinear regression analysis, and Eadie-Hofstee plots are used to evaluate the presence of a two-carrier transport system. Figure 1

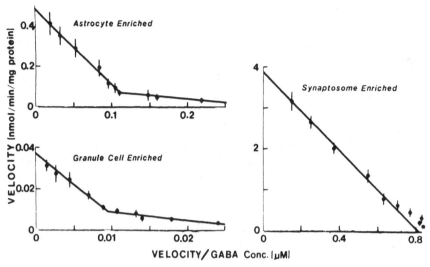

Fig. 1. Graphic representation (Eadie-Hofstee plot) of GABA uptake as a function of concentration into different cellular fractions. The slope of the plot gives K_m, the intersection of the line with the ordinate is V_{max}, and the intersection with the abscissa V_{max}/K_m. For astrocyte-enriched and granule cell-enriched fractions the Eadie-Hofstee plots of the uptake data appear to be nonlinear, indicating the uptake may be mediated by more than one transport system (from Shank and Campbell, 1984).

is a representation of Eadie-Hofstee plots illustrating preparations in which the data are nonlinear, indicating an uptake mediated by more than one carrier system and a preparation with linear data, i.e., a single-carrier system. In instances where evidence for two carriers is found, a regression analysis is performed with the program using rate law equations appropriate for uptake mediated by a single system and by two-carrier systems functioning independently. The computer program for the estimation of kinetic parameters in enzymatic rate laws is easy to learn and convenient to use. Copies are available through the Share Program Library Agency Catalog No. 360 D-.13.2.004.

3.3. Inhibitor Sensitivity

There are a number of inhibitors of GABA uptake known. Some of these are equally effective at inhibiting uptake in glial and neuronal cells, whereas others are more selective for one of these uptake processes. All of the known inhibitors are, however, taken up by the processes they inhibit, and via heteroexchange they release GABA, especially exogenously labeled GABA, from intracellular sites (Szerb, 1982). If endogenous GABA is measured, nipecotic acid can be used to analyse potassium-induced, calcium-dependent release of GABA by inhibiting its reuptake (Szerb, 1982). Nipecotic acid (K_i = 20 μM), guvacine (K_i = 25 μM), and homo-β-proline (K_i = 6–16 μM) are very potent inhibitors of GABA uptake, but are not selective for neuronal or glial uptake (Schousboe et al., 1983). Cis-3-aminocyclohexane carboxylic acid (K_i = 69 μM) selectively inhibits neuronal GABA uptake. Typical glial inhibitors are β-proline, homonipecotic acid, 4,5,6,7-tetrahydroisoxazalo[4,5c]pyridin-3-ol (THPO), and 5,6,7,8-tetrahydro-4H-isoxazolo[4,5c]azepin-3-ol (THAO). They are highly selective toward glial cells, but are not very potent inhibitors (Schousboe et al., 1979, 1981; Larsson et al., 1981, 1983). β-Alanine is a rather weak inhibitor of GABA uptake into cultured astrocytes. Studies with these inhibitors led to the suggestion that the two GABA carriers in glial and neuronal cells accept the GABA molecule in different conformations (Schousboe et al., 1983).

β-N-Oxalyl-L-α,β-diaminopropionate is a competitive inhibitor of the high-affinity uptake of glutamate into synaptosomes (Lakshmanan and Padmanaban, 1974). It is a neurotoxin from the seeds of *Lathyrus satidus*. Folate is an inhibitor of the uptake of glutamate and aspartate (McGeer et al., 1978), and D-aspartate has been found to be an inhibitor of astrocytic glutamate uptake (Schousboe et al., 1983).

Para-chloromercuriphenylsulfonate is the most potent glycine uptake inhibitor known to date, but imipramine, chlorpromazine, and hydrazinoacetic acid all have some inhibitory action (Aprison et al., 1970).

The uptake of taurine in glial cells and neurons is strongly inhibited in the presence of β-alanine. GABA and hypotaurine are strong inhibitors of taurine uptake in brain slices (Lahdesmaki and Oja, 1973). In cultured glioma cells, GABA and taurine are competitive inhibitors of each other's uptake, and this may indicate a common transport system.

3.4. Experimental Protocol

For estimating the initial rate of uptake of amino acids into cellular material, a detailed experimental protocol was selected from Shank and Campbell (1984) and is given below. Various other procedures that differ in detail, but contain the same concept, are to be found in the literature; for example, Iversen and Neal (1968).

In these experiments, 50 μL samples of cellular material containing the equivalent of 5–30 μg of protein are added to 150 μL of an incubation medium containing 0.02–0.05 μCi of a ^{14}C-labeled compound, or 0.1–0.2 μCi of a ^{3}H-labeled compound. Unless otherwise specified, the incubation medium is composed of NaCl (120 mM), KH$_2$PO$_4$ (3 mM), MgCl$_2$ (2 mM), CaCl$_2$ (2 mM), and glucose (5 mM), and is buffered with NaHCO$_3$ (24 mM). Prior to incubation, the medium is aerated with a mixture of O$_2$:CO$_2$ gas (95:5). The pH of the medium is 7.3–7.4. Specific concentrations of each compound studied are achieved by adding appropriate amounts of the nonlabeled form. The samples are incubated in polyethylene microcentrifuge tubes (400 μL) at 35°C. The period of incubation is usually 2, 4, 6, or 10 min, but in preliminary experiments, samples are incubated for as long as 20 min in order to establish the time interval over which accumulation of the radiolabel is linear. For blanks, samples are incubated at 0–2°C, or an amount of the nonradiolabeled form of the test compound is added to the medium to make a total concentration of 10 mM. Incubation is terminated by transferring the samples to an ice bath. The suspension of cellular material is then pelleted by centrifugation, the supernatant is removed by suction, and the surface of the pellet is washed twice with cold medium. The bottom 2–4 mm portion of the microcentrifuge tube that contains the pellet is cut off, blotted, and placed into a scintillation vial for determination of radioactivity. Radioactivity is determined by scintillation counting at approximately 90 and 35% efficiency, for ^{14}C and ^{3}H, respectively.

4. Release

Release of an amino acid may be studied by loading the preparation with radiolabeled, exogenous amino acids. Then the preparation is continuously superfused and the radioactivity in the collecting fluid is monitored. Alternatively, endogenous, nonlabeled amino acids released by the preparation can be determined in the superfusate.

4.1. Spontaneous Release

There is a spontaneous release of endogenous amino acids from neuronal and glial elements. To study the rate of this release and its possible stimulation by releasing agents, efflux experiments using desaturation curves from superfusion experiments are usually used (Winegard and Shanes, 1962; Cutler et al., 1971). Preparations (tissue slices or cellular material) are, after a certain period of preincubation (usually 10 min) in artificial solution, exposed to a solution containing the radiolabeled amino acid (0.6 μCi/mL for

[14]C-labeled compounds; 2 μCi/mL for ^{3}H-labeled compounds) for 30 min. Thereafter the material is transferred to a superfusion apparatus (Cutler et al., 1971) and superfused with nonradioactive solution. The effluent is collected into continuously changing vials. At the end of the superfusion period, both the radioactivity of the collection vials and the residual radioactivity in the tissue are determined. The initial content of labeled solute in the slices is obtained from the sum of residual tissue radioactivity and effluent radioactivity, and is expressed as 100%. The percentage of labeled solute remaining in the tissue at any time is calculated by subtracting the cumulative radioactivity recovered in the effluent from the initial tissue content (*see* Fig. 2 for an efflux time course).

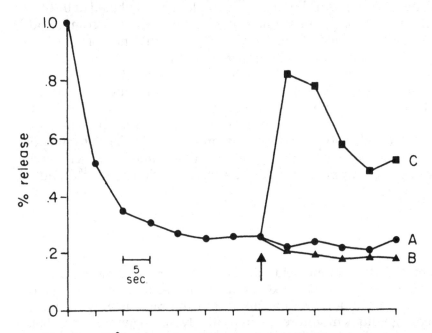

Fig. 2. Ca^{2+}-dependent K^+-stimulated GABA release from rat brain synaptosomes. A synaptosomal fraction is incubated with [^{14}C]GABA for 10 min at 37°C, then placed on the filter and perfused at 12 mL/min. Effluent is collected during 5-s intervals. Samples are first perfused with control medium containing 56 mM K^+ and no Ca^{2+}. After 40 s of perfusion, samples are treated as follows: (a) perfusion with control medium, plus 56 mM K^+ is continued; (b) sample is moved under a second syringe outlet and perfused with identical buffer solution (control medium with 56 mM K^+); (c) sample is moved under a third syringe that contains high K^+ medium and 3 mM Ca^{2+}. Percent release is calculated as the percent of radioactivity collected in each fraction relative to the total radioactivity present on the filter prior to collection of that fraction (from Redburn et al., 1975).

The semilogarithmic plot is resolved into components by the method of subtraction, and the half-time ($t_{1/2}$) of each component is determined graphically. The rate of efflux of labeled solute (k, %/min) is calculated from: $k = 0.693/t_{1/2} \times 100$ (Cutler et al., 1971). This method reveals a fast component for amino acids, as well as other compounds of different molecular weight. It has a $t_{1/2}$ of 2–3 min in tissue slices for all components studied and probably represents the washout of adherent medium. The slow component is the loss of isotope from the tissue as a whole; it is different for different substances and is sensitive to temperature changes. The apparent first order kinetics of the slow component does not preclude the possibility that the amino acids are lost from different tissue compartments at different rates. This does, however, suggest that the loss from one major compartment to another may be rate-limiting for clearance from the whole tissue (Cutler et al., 1971). In brain slices, the slow exponential loss of amino acids is linear throughout 40 min of superfusion, and at the end of this time, 60–70% of the labeled amino acids are recovered in the effluent (Cutler et al., 1971).

4.2. Use of Agents for Evoked Release

To investigate the possibility that an amino acid may be released as a transmitter, one has to monitor efflux from a preparation and then introduce depolarizing stimuli to simulate normal electrical activity. The evoked release that occurs during such treatments can be analyzed for calcium dependency and other ionic or metabolic requirements. Several agents may be used to induce transmitter release and are described below.

4.2.1. Ionophores

4.2.1.1. VOLTAGE-SENSITIVE SODIUM CHANNELS.   Veratridine, veratrin, protoveratrines, and batrachotoxin can be used to open voltage-sensitive (tetrodotoxin-sensitive) sodium channels, thereby depolarizing the neuronal membrane (Ulbricht, 1969). Glial membranes are not affected by these drugs (Tang et al., 1979) and, therefore, such drug treatment represents a way of selectively depolarizing neuronal cell membranes. It has been shown that treatment with protoveratrine enhances release of endogenous glutamate, GABA, aspartate, and glycine in brain slices, and that this release is suppressed by tetrodotoxin (Benjamin and Quastel, 1972). The opening of sodium channels will cause an increase in sodium influx and consequently an increase in cytoplasmic concentration of the ion. This increased internal sodium concentration will result in an enhanced release of those amino acids that are released in a sodium-dependent way. Calcium will enter neurons after depolarization through voltage-sensitive calcium channels (Baker and Glitsch, 1975). In veratridine-treated cells, the possibility exists that additional calcium enters the cytoplasm because the large increase in sodium alters the equilibria of intracellular calcium-sequestering mechanisms (Donatsch et al., 1977).

4.2.1.2. CALCIUM CHANNELS. The antibiotic A-23187 opens specific calcium channels in cell membranes of neurons and glial cells. In neurons, an increase in intracellular calcium leads to permeability changes for other ions (Latorre et al., 1984), with a resultant complex pattern of effects. Therefore, A-23187 may not be used as an agent that changes specific parameters in distinct cell types. However, it is known that A-23187 causes the release of amino acid transmitter substances (Salceda and Pasantes-Morales, 1975) in the retina. It is, however, not very effective in brain slices in this respect. This may be explained by insufficient diffusion into the tissue (Vargas et al., 1976). Also, cell damage after application of A-23187 has been reported (Chandler and Williams, 1977; Schlaepfer, 1977).

4.2.2. Electrical Stimulation

Certain biological membranes may be stimulated by the application of electrical pulses. When a certain threshold current density across such a membrane is reached, voltage-dependent ion channels open and this usually leads to a depolarization of the membrane. Such stimuli simulate the actions of invading action potentials. This technique may be applied to the study of depolarization-evoked release of amino acids. Work involving electrical stimulation has been carried out with brain slices and, to a lesser extent, with synaptosomal preparations (Bradford, 1970; De Belleroche and Bradford, 1972). Electrical current that passes through a biological preparation is normally short-circuited through the bulk of the external fluid, only a very small part of it crossing the biological membranes; it is this part that is the actual stimulating current. This current cannot be measured—only the total current passing through the preparation can be measured. Orrego (1979) suggests the use of tissue slices with the same thickness and deployment of a stimulating electrical field perpendicular to the slice surface, as is the case with quick-transfer electrodes. Under these conditions the electrical resistance per unit of slice area (i.e., resistivity) is approximately constant throughout a series of experiments with different slices. McIlwain and

Rodnight (1962) suggest the use of pure silver, gold, tungsten, or molybdenum as appropriate electrode materials in order to avoid electrolytic artifacts. Stimulating pulses that are alternating in polarity, with both phases perfectly balanced, represent suitable means of avoiding the formation of electrolytical products and intolerable increases in electrode resistance that may occur under some conditions (Lilly, 1961; Orrego and Miranda, 1976). Rectangular pulses may lead to electrochemical degradation of, for example, labeled glutamate (Haschke and Heavner, 1976). Therefore sine-wave currents should be used (McIlwain, 1975). When 50 Hz sine-wave current is used as a stimulus, amino acid release shows a characteristic "voltage profile" (Orrego et al., 1974). Up to 1.5 V, substances are released only in a calcium-dependent manner and this release can be blocked by tetrodotoxin, indicating that this is the voltage range in which physiological stimulation and, therefore, release of transmitter substances occurs. With voltages higher than 3 V, a nonspecific calcium-independent release, probably caused by "nonphysiological" stimulation, occurs. Endogenous GABA is released with physiological stimuli, but exogenously applied radiolabeled GABA is released only with "nonphysiological" stimuli (Valdes and Orrego, 1978; Orrego and Miranda, 1976; Orrego et al., 1976); this is probably a problem of metabolic compartmentation (see above).

4.2.3. High Extracellular Potassium Concentration

Evoked release by high potassium concentrations is preferred by most authors. For synaptosomes, concentrations used are 10–55 mM potassium, a range in which the amount of released GABA is proportional to the potassium concentration (Cotman et al., 1976). The degree of release has been correlated to the estimated membrane potential of the synaptosomes. For brain slices, potassium concentrations of at least 40 mM are needed for release of GABA and aspartate, 28 mM for glutamate release, and 14 mM for glycine release. The release is calcium-dependent, which suggests that it is caused by stimulation within the physiological range. In this regard, Orrego (1979) proposed that potassium would release amino acids mainly from a cytoplasmic pool, rather than from vesicles, and that calcium dependence must be used as an additional criterion for release (see below). There are also conflicting reports about the cellular site of release (see below).

4.3. Calcium Dependence of the Release

Excitation–secretion coupling is a highly calcium-dependent process, and thus the calcium dependency of evoked release could be one measure with which to demonstrate this mechanism. Unfortunately, calcium influences other parameters, such as potassium and sodium fluxes and cell volume *per se* (Hoffmann et al., 1984). For this reason, Vargas and Orrego (1976) introduced the terms "true" and "pseudo" calcium-dependency. They studied release of a vesicular neurotransmitter (noradrenaline) and a cytoplasmic amino acid (α-aminoisobutyrate) from brain cortex slices, and found calcium dependence of the evoked release with "true" dependency to be up to 1.5–2 m*M* calcium, and with "pseudo" dependency, up to 200 m*M* calcium. They suggest that any study on evoked release should include the criterion of calcium dependency. In the lower range (up to 2 m*M*) it would represent "true" calcium dependency and, therefore, release from synaptic vesicles. A dependence on much higher levels of extracellular calcium would be indicative of a "pseudo" calcium dependency and, therefore, a cytoplasmic location of the amino acid.

4.4. Selected Methodological Problems

Below are some major problems common to release studies with all preparations described.

4.4.1. Cellular Site of the Release

Synaptosomes and brain slice preparations (Cotman et al., 1971; Henn et al., 1976; Morgan, 1976) consist of heterogenous cell material, i.e., at least two different cell types–neuronal and glial cells. Glial cells show spontaneous release of amino acids, and the critical question is whether or not some of the evoked release represents increased glial, rather than neuronal, release.

To address this question, several experimental strategies can be adopted. One strategy takes advantage of the fact that in some preparations, such as dorsal root ganglia, sympathetic and spinal ganglia, and rat retina, only glial cells take up exogenous amino acids or, in the latter cases, exogenous GABA (Minchin and Iversen, 1974; Neal and Bowery, 1979; Bowery et al., 1979). GABA is preferentially taken up into cortical slices and frog retina. If the subsequent release of radiolabeled compounds is studied, one may obtain information about the localization and differences in the mechanism of release of different cell types. The problems associated with differential labeling of releasable transmitter pools, of course, remain. Veratridine-like drugs act as releasing agents for neurons, but not for glial cells (see above), and their application to heterogenous cell material can lead to selective release of neuronal amino acids. Cultures enriched in neurons or glial cells have been used to investigate release (Pearce et al., 1981), as have homogeneous primary cultures of astrocytes (Hertz et al., 1978; Schouseboe et al., 1983) and bulk-prepared glial cells (Sellstrom and Hamberger, 1977). Most experiments suggest that the calcium dependence of amino acid release evoked by high potassium has a neuronal origin.

Another question concerns the vesicular or cytoplasmic origin of amino acids released from neurons. From investigations of the relative action of noradrenaline (which is thought to be stored exclusively in vesicles) and α-aminoisobutyric acid (AIB) (which is taken up into cells, but is not metabolized and remains in the cytoplasm), Vargas et al. (1977a,b) postulated that amino acids that can be released by electrical stimulation of up to 3 V and have an absolute calcium requirement are released from synaptic vesicles. If release evoked by high potassium is analyzed, it is necessary to determine the true time course of the release. Amino acids that are released with an efflux peak occurring within 2–4 min of potassium elevation and demonstrate a diminishing rate of release despite the continued presence of high potassium, are considered to be released from synaptic vesicles; they have an absolute potassium requirement for their release. Amino acids that attain their maximal release rates after only 10 min following potassium elevation, however, and maintain these rates without decrement, are considered to be of cytoplasmic origin.

4.4.2. Problems of Compartmentation That Arise from the Use of Exogenous Compounds

The use of exogenous compounds (radiolabeled amino acids or their radiolabeled precursors) raises several problems. First, one has to ensure that the radiolabeled compound is indeed accumulated by neuronal elements in vitro. Second, multiple neuronal pools for GABA and possibly other amino acids exist, and it is now established that short-term exposure of a preparation to the radiolabeled compound does not label each of these pools to an equivalent extent (Ryan and Roskoski, 1975; Levy et al., 1976; Gauchy et al., 1977). It seems certain, therefore, that exogenous compounds do *not* enter the same releasable pools as their endogenous counterparts. It is clear that in the case of GABA (for which more detailed studies are available), newly captured and newly synthesized GABA is released in preference to the endogenous GABA store. Gauchy et al. (1977) found that there is also a difference between newly synthesized and newly accumulated GABA. The newly synthesized GABA is released from a smaller pool exhibiting more rapid turnover than newly accumulated GABA. On the other hand, Haycock et al. (1978) found that newly accumulated GABA is released in preference to endogenous GABA only under calcium-free conditions. In normal solutions, release of both pools are similar. It was found that introduction of calcium into the superfusion medium results in an increased efflux of radiolabeled GABA that returns to basal levels after 30 s. This decrease is not associated with a depletion of radiolabeled GABA. The endogenous compound displays more sustained levels of release and this methodology was recommended by the authors as more suitable for the study of evoked GABA release. In the case of glutamate and aspartate, in response to an increase in external calcium only about 50% of the released radiolabel was associated with these amino acids. As already pointed out in section 3.1.1., AOAA can be used successfully to block degradation of exogenous GABA during release studies. When endogenous amino acid release is studied, there is no problem with differential compartmentation, but labeled amino acids may be used to study the origin of the stimulus-evoked release of amino acid, i.e., vesicular vs cytoplasmic (the cytoplasmic pool exhibits a higher turnover than does stored GABA). The use of endogenous amino acids has to be recommended in all cases in which one is interested in studying the occurrence and properties of the stimulus-evoked release of amino acids.

4.5. Experimental Protocol

An experimental protocol is described below for monitoring efflux rates of amino acids from cellular preparations. The methodology can be used for a variety of preparations, including homogenates, subcellular fractions, small tissue slices, and cell suspensions. The experimental protocol given here is for studying [^{14}C]-GABA efflux from rat brain synaptosomal fractions, as described by Cotman and coworkers (Levy et al., 1973; Redburn et al., 1975; Cotman et al., 1976).

Crude mitochondrial fractions containing synaptosomes are prepared from rat forebrains by the method of Cotman and Matthews (1971). The brains are homogenized in cold sucrose (320 mM) in a Teflon—glass homogenizer. A crude nuclear fraction is removed by centrifugation at 5500 rpm (2200g) in a Spinco 30 rotor for 2 min. The resultant supernatant is centrifuged at 13,000 rpm (14,000g) for 12 min. The crude mitochondrial fraction (P$_2$) thus formed is washed in sucrose (320 mM) and again pelleted as the crude mitochondrial fraction (P'$_2$). About 45% of the particles in this fraction are synaptosomes, as determined by electron microscopic examination. This fraction is termed the synaptosomal fraction. The preparations are resuspended in an incubation solution consisting of 150 mM NaCl, 20 mM HEPES, 10 mM glucose, 5 mM KOH, 1.2 mM Na_2HPO_4, 1.2 mM $MgSO_4$, 0.1 mM AOAA, 1 mM ascorbic acid, and buffered to pH 7.4 with Tris base. They are pelleted and kept in pellet form at 0–4°C until incubation. The tissue can be stored as the pellet for up to 2 h at 0–4°C without apparent change in release characteristics. Tissue pellets are resuspended in incubation solution to give a protein concentration of 0.6–0.9 mg/mL and incubated for 10 min at 37°C under hydrated O_2:CO_2 (95:5%) with 0.44 μM [^{14}C]-GABA. Immediately following incubation, aliquots (usually 1 mL) of the tissue suspension are placed on four filters. Following suction of the tissue suspension, seven 2-mL washes of calcium-deficient wash solution are applied to the filters. The wash solution has the following composition: 150 mM NaCl, 20 mM HEPES, 10 mM glucose, 5 mM KOH, 5–50 mM KCl (variable), 1 mM $MgCl_2$, 0–1 mM AOAA, 1 mM ascorbic acid, and Tris base to make the pH 7.4. Wash solution is applied for 30 s every 60 s. Washout in the presence of elevated potassium is stabilized by the seventh wash (*see* Fig. 2). Stimulating alkaline earth cations (calcium) are introduced only in the seventh wash (Fig. 2). Investigation of the time course of release stimulation is essentially the same as above, except that filters with previously incubated and plated tissue are placed in a positive-pressure perfusion apparatus (*see* Fig. 3) and samples are collected every 5 s or 250 ms. All release measurements are conducted at room temperature (22–24°C). One mL of wash filtrate is counted in 10 mL Triton X-100/toluene/phosphor medium (Patterson and Greene, 1965) and corrected for quenching. In the presence of AOAA, 95% of the [^{14}C] has been previously demonstrated to be associated with GABA (Levy et al., 1973). Radioactivity in the filters is counted in the same manner in 20 mL of the same medium following overnight solubilization in 2 mL of a 1% solution of sodium dodecyl sulfate containing 20 mM EDTA and adjusted to pH 8 with NaOH. Efflux for a given collection period is expressed as that percentage of the total radioactivity present on the filters immediately before the wash of interest. Importantly, release (measured from the seventh wash) is expressed as the difference between efflux from simultaneously run filters treated identically except for the additional cation, as shown in Eq. (1).

$$\text{Inhibition of release} = (R_N - R_I)/R_N \times 100$$

where R_N = release without inhibitor, and R_I = release with inhibitor. This expression of the data assures that calcium-dependent release is analyzed independently of potassium-dependent effects.

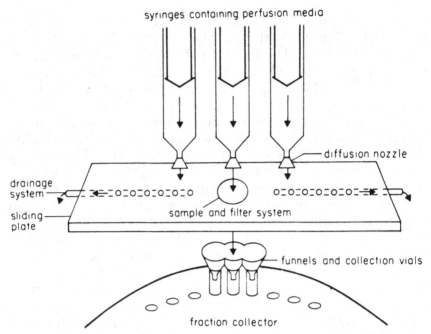

Fig. 3.   Schematic diagram showing major components of the perfusion apparatus described by Redburn et al. (1975). The sample is retained on a filter and perfused with medium from any one of the three
syringes. Diffusion nozzles spread the stream of medium emerging
from each syringe so that it will flow evenly over the surface of the filter.
Drops of perfusate emerging from the sample are collected in funnels
that direct the perfusate into collection vials held in a fraction collector.
Perfusion medium is changed by moving the sample from under the
perfusion stream of one syringe to another. Syringes not open to the
sample empty into a drainage system (from Redburn et al., 1975).

5. Preparations Used in the Study of Amino Acid Transport

The following section of this review presents a list with short de-
scription of the major preparations that are used to study amino
acid transport. Their limitations are discussed to make the potential user aware of possible artifacts and pitfalls. Ideally, a combination of two or more different preparations should be used to
strengthen observations and allow more general conclusions to be
drawn, as carried out, for example, by Bradford (1981).

5.1. Synaptosomes

Synaptosomal preparations are the most widely used model systems in studying amino acid transport of the brain. The relevance

of synaptosomal preparations for transport processes has been reviewed by Levi and Raiteri (1976). Care must be taken concerning which fraction is used for uptake or release studies. Crude mitochondrial fractions (P_2) accumulate GABA and glutamate, reflecting a mitochondrial property (Levi et al., 1974). Homogenates contain large amounts of free endogenous compounds released during homogenization, and these may interfere directly or indirectly with the transport of various substrates (Levi and Raiteri, 1976). Purified synaptosomes probably represent the most appropriate fraction when uptake and release in nerve endings are to be tested.

There are also problems with endogenous substances present in incubation media, because of leakage from the tissue, when high ratios between amount of tissue and volume of medium are used. This is a problem when radioactive substrates are used at low concentrations. The problems of loss of amino acids by cold-shock treatment (a necessary step in stopping the transport reaction) and homoexchange of amino acids are probably major obstacles and sources of artifacts with this preparation. These factors have been discussed separately above. To study amino acid transport, which is sensitive to the electrochemical driving force across the membrane, it is important that membrane damage is minimized in order to maintain a normal membrane potential. Damaged synaptosomal membranes might contribute to several of the differences in amino acid transport observed with synaptosomes, brain slices, and intact cells. In discussing this problem, Pastuszko et al. (1981) have pointed out that the isolation of synaptosomes in Ficoll density gradients, instead of sucrose density gradients, yields preparations of better quality. Filtration through millipore filters may also cause damage to synaptosomal membranes, so Pastuszko et al. (1981) used a method involving rapid centrifugation through silicone oil.

5.2. Tissue Slices

Tissue slices have a certain advantage over synaptosomes: They constitute a tissue sample with more intact spatial relations and therefore probably less chances of membrane damage. A disadvantage is that of cellular heterogeneity, as discussed in section 4.4.1. A useful introduction to the methodology of brain slices was recently published by Schwartzkroin (1981). A single tissue slice may be used in an experiment for 8–10 h. A major advantage is that amino acid release can be evoked by stimulation of intact nerve tracts (Collins, 1974) and this preparation proved essential for the provision of evidence for the transmitter function of certain amino acids. A major problem is the limitation imposed on rapid replacement of extracellular fluid. Another disadvantage is swelling of brain slices when they are incubated in vitro (Pappius and Elliott, 1956). However, Cohen (1974) has presented evidence for the independence of uptake processes from tissue swelling in vitro.

The thickness of slices obviously has an impact on the accumulation of amino acids (Sershen and Lajtha, 1974). Glutamate, aspartate, and GABA are accumulated to a greater extent in thinner (0.1 mm) slices than in thicker (0.42 mm) slices. Glutamate seems to accumulate at the surface of thick slices, and thin slices may be damaged to a greater extent than thick ones (Cohen, 1974).

Glutamate, aspartate, and GABA may be used as substrates for the rapidly turning tricarboxylic acid cycle because mitochondrial oxidative phosphorylation may be uncoupled in damaged tissue (Fagg and Lane, 1979). Fagg and Lane (1979) have suggested that, because of these difficulties, data from synaptosome fractions and homogenous cell cultures may provide more useful information.

5.3. Brain Tissue In Vivo

Uptake studies in vivo involve mainly autoradiographic methods and are discussed in detail in another chapter of this volume (Scott-Young, 1985). For the acute and chronic collection of released amino acids in vivo, several methods have been developed and are described briefly.

5.3.1. Cortical Cup Technique

The cups are small plastic cylinders that are used to investigate the release of amino acids from cortical structures in acute and chronic experiments. For acute experiments, Clark and Collins (1976) used a Perspex cup, 4 mm id, that was placed on the surface of the visual cortex. The junction between the brain and cup was sealed with soft white paraffin. The cup contained 50 μL Ringer solution and was exchanged every 10 min (*see* Fig. 4). The ion composition of the solution could be changed to study effects on the release. The amino acids in the "exchanged" solution were determined and the temperature in the cup and the electrocorticogram of the surface in the cup were monitored. A bipolar platinum electrode was inserted 0.5 mm below the surface of the cortex either within or outside the cup for electrical stimulation. During these experiments the animal was anesthetized.

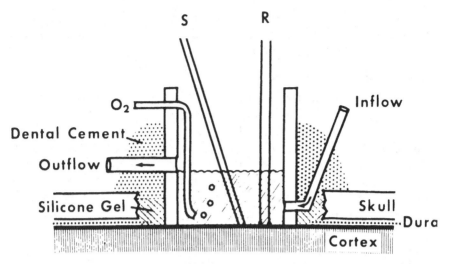

Fig. 4. Diagram of a cortical cup used by Jasper and Koyama to collect amino acids from the cortex surface. The cup is fixed in the skull with dental cement and sealed to the surface of the cortex with silicone gel. The inside diameter of the cup is approximately 1.1 cm, covering a surface of 1 cm^2 (from Jasper and Koyama, 1969).

A further application of the above technique was developed by Dodd et al. (1974) and Dodd and Bradford (1974, 1976) to monitor amino acid release over periods of days from the surface of the cortex in unanesthetized, unrestrained rats with a superfusion cannula implanted in the skull. Using this technique, animals can be studied in the absence of effects of physiological shock or anesthesia, and in the presence or absence of drugs. Fourteen μL of solution were in direct contact with the cortex; the flow-rate was 5 mL/h, and the fluid exchange was under sterile conditions. Four-hour fractions were collected. Diurnal variation in amino acid release and release during experimentally induced epilepsy were investigated. In another study electrodes were implanted to allow activation of sensorimotor cortex, and chemical depolarizing agents were introduced into the superfusion stream (Bradford, 1981). Drugs were applied systemically or via cannulae.

5.3.2. Perfusion of Cerebral Fluid Spaces

To study other regions of the CNS, cerebrospinal fluid spaces have been perfused. Fagg et al. (1978) studied the release of endogenous amino acids into the perfused central canal of the cat spinal cord. A hypodermic needle was lowered into the central canal and cannulated at the sacral end with polyethylene tubing. Warmed artificial CSF was perfused (60 μL/min) through the canal and samples of the perfusate were collected every 10 min (Morton et al., 1977). Descending spinal tracts were electrically activated with a bipolar platinum electrode.

5.3.3. Push–Pull Cannulae

Reubi and Cuenod (1976) studied the effect of electrical stimulation on release of labeled substances previously injected into the tectum of the pigeon. The animals were anesthetized. Radiolabeled amino acids were injected at a depth of 800 μm into the tectum with a microsyringe driven by a motor (0.2 μL/min). The needle was withdrawn after injection. In order to collect the radioactive material released from the tectal cells, a push–pull cannula was inserted into the tectum at the end of the injection period. This system consisted of two stainless-steel tubes with an inner diameter of 0.2 mm and an outer diameter of 0.3 mm. The inflow tube was beveled to an angle of 45° and was 0.5 mm longer than the nonbeveled outflow tube. Inflow and outflow (40 μL/min) were coupled by a peristaltic pump and a recording electrode was attached to the cannula. Vials, which collected the outflow, were changed every 3 min for the measurement of radioactivity. Outflow was monitored during electrical stimulation by an electrode inserted into either the nuclear isthmi pars parvocellularis or the optic nerve papilla.

5.3.4. Critique

Investigations using these three techniques yielded evidence for a role of aspartate, glutamate, GABA, and glycine as neurotransmitters. To date, these in vivo methodologies present an alternative to in vitro cell systems. However, some disadvantages were pointed out by Fagg and Lane (1979). One problem is that it is not really possible to introduce the calcium-dependence criterion into these experiments, since the intact tissue cannot really be depleted of its calcium levels. The release of amino acids, therefore, does not necessarily represent transmitter release; it may also be a result of metabolic changes in the stimulated tissue. Most collections have been made from surfaces of tissue interfaces and probably do not allow any conclusions concerning the situation in deeper cell layers. Most importantly, the push–pull cannula technique may result in intolerable tissue damage at the cannula tip (Chase and Kopin, 1968).

5.4. Cell and Tissue Culture

Explant and aggregated cell cultures consist of heterogenous cell material and have been used for autoradiographic studies (Hosli and Hosli, 1976; Lasher, 1974). In order to conduct transport studies it is necessary to have homogeneous cell material. Neuronal and glial cell lines have been used to study amino acid transport (Hutchison et al., 1974; Schrier and Thompson, 1974), but there are serious problems about the normality of these cells regarding crucial parameters like membrane potential (Hertz et al., 1985). Astrocytes in primary cultures have been used to investigate amino acid transport in glial cells (Hertz et al., 1978; Schousboe et al., 1983; Ramaharobandro et al., 1982) and the biochemical and functional characterization of these cultures has suggested that cultured astrocytes may be well suited as a model system for their in vivo counterparts (Hertz, 1982). Recently it has been possible to selectively culture glutamatergic and GABAergic neurons. Hertz et al. (1980) and Yu and Hertz (1982) have chosen cultured cerebellar granule cells as a model system for glutametergic neurons for the study of amino acid transport. Cultured cerebral cortical interneurons (stellate cells) were studied as a model of a GABAergic neuronal population (Larsson et al., 1981, 1983; Yu and Hertz, 1982). These cultures all consist of monolayers attached to the culture dish, and transport studies are therefore relatively straightforward, entailing no problems during exchange of solutions or during the washing procedure. There are no problems with damaged cell membranes. Pearce et al. (1981) described a simple and inexpensive chamber which, because of its novel inlet and outlet manifold systems, ensures a rapid exchange of perfusion solutions with minimal disruption of the cultures. Recent improvements in the preparation and culturing techniques have reduced problems associated with the contamination with other cell types (Hertz et al., 1985). With these preparations the only major concern is the extent to which the culturing process and the interruption of spatial relations change the normal properties of cells in primary cultures. Results obtained only with primary cultures and not with other model systems must therefore be questioned.

5.5. Isolated Ganglia and Retina

The rat dorsal root ganglion (Minchin and Iversen, 1974), spinal ganglia (Roberts, 1974), and superior cervical ganglion (Bowery and Brown, 1972) represent *in situ* systems with certain advantages. They can be used for the analysis of some aspects of GABA uptake and release. Radiolabeled GABA is exclusively accumulated into glial cells. Thus, these preparations consist of in situ systems with normal connections intact and, therefore, little chance of cell damage. They may be used to study the uptake and release of GABA into glial cells. It should be noted that this privilege of exclusive uptake into glial cells does not extend to the other amino acids that are accumulated by both neurons and glial cells (Roberts, 1974). Similarly, in the isolated retinae of rats, GABA is taken up predominantly by glial cells. In contrast, in the frog retina GABA is predominantly taken up into neuronal elements (Lake and Voaden, 1976). Thus, some isolated ganglionic and retinal preparations may be used selectively to study the properties of GABA transport into different cell types.

6. Experiments With Highly Purified Amino Acid Carrier Preparations

Kanner (1978a) solubilized the GABA carrier from synaptosomal fractions from rat brain and incorporated it into liposomes. This allowed direct flux experiments with the GABA carrier. The main features of carrier-mediated GABA transport in native membrane vesicles, such as the dependence upon sodium ions, chloride ions, and ion gradients, as well as electrogenicity and inhibitor sensitivity, were preserved in the reconstituted system. This seems to be a new and important direction for studying the transport of amino acids. It is very likely that this methodology will develop into one of the most important ways of studying amino acid transport systems and, therefore, the experimental protocol from Kanner (1978a) is given.

Synaptosomal fractions are isolated from rat brain and purified using Ficoll gradient centrifugation (Kanner, 1978b). Membrane vesicles are obtained after osmotic shock of either the 2–8% interfaces and/or the 8–12% interfaces. Aliquots are stored in liquid air.

For solubilization of the GABA carrier, these stored membrane vesicles are used and all subsequent steps are performed at 0–4°C. Membrane vesicles (20–50 mg protein) are incubated at a concentration of 2 mg/ml with 0.32M sucrose, buffer (10 mM Tris-sulfate, pH 7.4, 0.5 mM EDTA, 1 mM MgSO$_4$), 1 mM dithiothreitol, and 0.5% Triton X-100 (last addition). After 2 min of stirring, the mixture is centrifuged in a Beckman 50-Ti rotor for 1 h at 46,000 rpm. Biobeads SM-2 (Holloway, 1973), 0.3 g/mL, are added to the supernatant and the mixture is stirred for 2 h. The mixture is then centrifuged for 5 min at 10,000 rpm in a Sorvall SS-34 rotor to remove the Biobeads. The supernatant is brought to 50% saturation by addition of an equal volume of ammonium sulfate. After

10 min stirring, the mixture is centrifuged in the SS-34 rotor at 15,000 rpm for 20 min. The floating pellet is collected and resuspended at 5–10 mg/protein/mL in a solution containing 50 mM sucrose and 0.5 mM dithiothreitol in buffer. Aliquots are stored in liquid air. Under these conditions the protein can be stored for at least 2 mo, without loss in transport activity, and assayed after reconstitution.

Reconstitution is performed using a variant of the cholate dialysis procedure (Kagawara and Racker, 1971). Soybean phospholipids are dried under a stream of nitrogen and suspended, using a bath type sonicator, at 30 μmol/mL in dialysis buffer (see below) containing 33 mM octylglucoside. This phospholipid suspension, 18–20 μmol/mL, is incubated with the solubilzed protein, 1 mg/mL in a solution containing 50 mM potassium phosphate, pH 6.8, 1 mM dithiothreitol, 0.32M sucrose, buffer and 0.8% cholic acid neutralized with NaOH (last addition). This mixture is dialyzed for 18–20 h against 120–200 vol. dialysis buffer, which contains 120 mM potassium phosphate, pH 6.8, 0.32M sucrose, buffer, 1 mM dithiothreitol, and 1% glycerol. The proteoliposomes obtained after dialysis are used for the transport experiments.

Transport of GABA is measured by adding 15–40 μL of proteoliposomes to 0.36 mL of an incubation mixture containing 0.15M NaCl, 2 mM MgSO$_4$, and 0.07–0.28 μM [^{3}H]-GABA at 3.5–35 Ci/mmol. After incubation for various times at room temperature (20–23°C), the reactions are terminated by addition of 2 mL ice cold incubation mixture [not containing (^{3}H)-GABA] and filtration through membrane filters (Millipore, 0.22 μm pore size). After washing of the filters with another 2 mL of the above solution, the retained radioactivity is determined using liquid scintillation spectrometry. Zero times are obtained by adding the stopping solution prior to the proteoliposomes. Experimental values are up to 10–30 times the zero time values. NaCl is used at a concentration of 0.15M instead of the 0.1M used for the native vesicles (Kanner, 1978b), since with the former salt concentration 50% higher values in that system are obtained.

7. Conclusion

This discussion of the important and most used methodologies for uptake and release studies of amino acid neurotransmitters reveals a few guidelines every researcher starting to undertake research in amino acid-mediated neurotransmission should consider. It is advisable to use in almost any case endogenous amino acids rather than exogenous ones, although it might involve a more sophisticated analytical procedure. At least a comparison between both methodologies should be attempted. Regarding the uncertainties of all results obtained with the different biological preparations, one should use two or even more different preparations to strengthen observations. This is especially important for the release studies. Some future trends are visible: It will become more and more important to isolate the carriers involved in uptake and release and to study them separately in artificial membrane systems. This will allow the collection of data concerning the underlying basic transport mechanisms with less errors than in biological preparations.

Acknowledgments

The author is a scholar of the Medical Research Council of Canada. Thanks to Dr. Andrew J. Greenshaw for helpful suggestions in improving the text.

References

Aprison M. H., Davidoff R. A., and Werman R. (1970) Glycine: Its Metabolic and Possible Roles in Nervous Tissue, in *Handbook of Neurochemistry*, Vol. 3, (Lajtha A., ed.) Plenum, New York, pp. 381–397.

Aprison M. H. and McBride W. J. (1973) Evidence for the net accumulation of glycine into a synaptosomal fraction isolated from the telencephalon and spinal cord of the rat. *Life Sci.* **7**, 583–590.

Baker P. F. and Glitsch H. G. (1975) Voltage-dependent changes in the permeability of nerve membranes to calcium and other divalent cations. *Phil, Trans. Roy. Soc.* (B) **270**, 389–409.

Beart P. M., Kelly J. S., and Schon F. (1974) γ-Aminobutyric acid uptake in the rat peripheral nervous system, pineal, and posterior pituitary. *Biochem. Soc. Trans.* **2**, 266–268.

Benjamin A. M. and Quastel J. H. (1972) Locations of amino acids in brain slices from the rat. Tetrodotoxin-sensitive release of amino acids. *Biochem. J.* **128**, 631–646.

Bowery N. G. and Brown D. A. (1972) Gamma-aminobutyric acid uptake by sympathetic ganglia. *Nature New Biol.* **238**, 89–91.

Bowery N. G., Brown D. A., Marsh S., Adams P. R., and Brown D. A. (1979) Gamma-aminobutyric acid efflux from sympathetic glial cells. Effect of depolarizing agents. *J. Physiol.* **293**, 75–101.

Bradford H. F. (1970) Metabolic response of synaptosomes to electrical stimulation: Release of amino acids. *Brain Res.* **19**, 239–247.

Bradford H. F. (1981) GABA release in vivo and in vitro: Responses to Physiological and Chemical Stimuli, in *Regulatory Mechanisms of Synaptic Transmission* (Tapia R. and Cotman R. W., eds.) Plenum, New York, pp. 103–140.

Chandler D. E. and Williams J. A. (1977) Intracellular uptake and α-amylase and lactate dehydrogenase releasing actions of the divalent cation ionophore A 23187 in dissociated pancreatic acinar cells. *J. Membr. Biol.* **32**, 201–230.

Chase T. N. and Kopin I. J. (1968) Stimulus-induced release of substances from olfactory tubercle using push–pull cannula. *Nature* (Lond.) **217**, 466–467.

Clark R. M. and Collins G. G. S. (1976) The release of endogenous amino acids from the rat visual cortex. *J. Physiol.* **262**, 383–400.

Cohen S. R. (1974) The dependence of water content and extracellular marker spaces of incubated mouse brain slices on thickness, alterations produced by slicing and fluid spaces in intact and altered tissue. *Exp. Brain. Res.* **20**, 435–457.

Collins G. G. S. (1974) The spontaneous and electrically evoked release of [^{3}H]-GABA from the isolated hemisected frog spinal cord. *Brain Res.* **66**, 121–137.

Cotman C. W., Haycock J. W., and White W. F. (1976) Stimulus-secretion coupling processes in brain: Analysis of noradrenaline and gamma-aminobutyric acid release. *J. Physiol.* **254**, 475–505.

Cotman C. W., Herschmann H., and Taylor D. (1971) Subcellular fractionation of cultured glial cells. *J. Neurobiol.* **2**, 169–180.

Cotman C. W. and Matthews D. A. (1971) Synaptic plasma membranes from rat brain synaptosomes: Isolation and partial characterization. *Biochem. Biophys. Acta* **249**, 380–394.

Cox D. W. G. and Bradford H. F. (1978) Uptake and Release of Excitatory Amino Acid Neurotransmitters, in *Kainic Acid as a Tool in Neurobiology* (McGeer E. G., ed.) Raven Press, New York, pp. 71–93.

Cutler R. W. P. Hammerstad J. F., Cornick L. R., and Murray J. E. (1971) Efflux of amino acid neurotransmitters from rat spinal cord slices. I. Factors influencing the spontaneous efflux of [^{14}C] glycine and ^{3}H-GABA. *Brain Res.* **35**, 337–355.

De Belleroche J. S. and Bradford H. F. (1972) Metabolism of beds of mammalian cortical synaptosomes: Response to depolarizing influences. *J. Neurochem.* **19**, 585–602.

Dodd D. R. and Bradford H. F. (1974) Release of amino acids from the chronically superfused mammalian cerebral cortex. *J. Neurochem.* **23**, 289–292.

Dodd D. R. and Bradford H. F. (1976) Release of amino acids from the maturing cobalt-induced epileptic focus. *Brain Res.* **111**, 377–388.

Dodd D. R., Pritchard M. J., Adams R. C. F., Bradford H. F., Hicks G., and Blanshard K. C. (1974) A method for the continuous, long-term superfusion of the cerebral cortex of unanesthetised, unrestrained rats. *J. Sci. Inst.* **7**, 897–901.

Donatsch P., Lowe D. A., Richardson B. P., and Taylor P. (1977) The functional significance of sodium channels in pancreatic beta-cell membranes. *J. Physiol.* **267,** 357–376.

Fagg G. E., Jones I. M., and Jordan C. C. (1978) Descending fiber-mediated release of endogenous glutamate and glycine from the perfused cat spinal cord in vivo. *Brain Res.* **158,** 159–170.

Fagg G. E. and Lane J. D. (1979) The uptake and release of putative amino acid neurotransmitters. *Neuroscience* **4,** 1015–1036.

Gauchy C. M., Iversen L. L., and Jessell T. M. (1977) The spontaneous and evoked release of newly synthesized [14] GABA from rat cortex in vitro. *Brain Res.* **138,** 374–379.

Haschke R. and Heavner E. (1976) Glutamate breakdown during electric field stimulation. *Brain Res.* **102,** 351–354.

Haycock J. W., Levy W. B., Denner L. A., and Cotman C. W. (1978) Effects of elevated $[K^+]_O$ on the release of neurotransmitters from cortical synaptosomes: efflux or secretion. *J. Neurochem.* **30,** 1113–1125.

Henn F. A. (1982) Neurotransmitters and Astroglia Lead to Neuromodulation, in *Chemical Transmission in the Brain*

(Bujs, R. M., Pevet P., and Schwab D. F. eds.) Elsevier, Amsterdam, pp 241–252.

Henn F. A., Anderson D. J., and Rustad D. G. (1976) Glial contamination of synaptosomal fractions. *Brain Res.* **101,** 341–344.

Hertz L. (1979) Functional interactions between neurons and astrocytes. I. Turnover and metabolism of putative amino acid transmitters. *Prog. Neurobiol.* **13,** 277–323.

Hertz L. (1982) Astrocytes, in *Handbook of Neurochemistry,* Vol. I, 2nd Edition (Lajtha A., ed.) Plenum, New York, pp. 319–355.

Hertz L., Juurlink B. H. J., Szuchet S., and Walz W. (1985) Cell and Tissue Cultures, in *Neuromethods,* Vol. 1 (Boulton A. A. and Baker G. B. eds.), Humana, Clifton, New Jersey.

Hertz L., Kvamme E., McGeer E., and Schousboe A. (1983) *Glutamine, Glutamate, and GABA in the Central Nervous System.* Alan R. Liss, New York.

Hertz L., Schousboe A., Boechler N., Mukerji S., and Fedoroff S. (1978) Kinetic characteristics of the glutamate uptake into normal astrocytes in culture. *Neurochem. Res.* **3,** 1–14.

Hertz L., Yu A., Svenneby G., Kvamme E., Fosmark H., and Schousboe A. (1980) Absence of preferential glutamine uptake into neurons. An indication of a net transfer of TCA constituents from nerve endings to astrocytes? *Neurosci. Lett.* **16,** 103–109.

Hoffman E. K., Simonsen L. O., and Lambert I. H. (1984) Volume-induced increase of K^+ and Cl^- permeabilities in Ehrlich ascites tumor cells. Role of internal calcium. *J. Membrane Biol.* **78,** 211–222.

Holloway P. W. (1973) A simple procedure for removal of triton X-100 from protein samples. *Anal. Biochem.* **53,** 304–308.

Hosli E. and Hosli L. (1976) Autoradiographic studies on the uptake of [3H] noradrenaline and [3H] GABA in cultured rat cerebellum. *Exp. Brain Res.* **26,** 319–324.

Hutchison H. T., Werrbach. K., Vance C., and Haber B. (1974) Uptake of neurotransmitters by clonal lines of astrocytoma and neuroblastoma in culture. I. Transport of γ-aminobutyric acid. *Brain Res.* **66,** 265–274.

Iversen L. L. and Neal M. J. (1968) The uptake of [^{3}H]-GABA by slices of rat cerebral cortex. *J. Neurochem.* **15,** 1141–1149.

Jasper H. H. and Koyama I. (1969) Rate of release of amino acids from the cerebral cortex in the cat as affected by brain stem and thalamic stimulation. *Can. J. Physiol. Pharmacol.* **47,** 889–905.

Johnstone R. M. (1979) Electrogenic amino acid transport. *Can. J. Physiol. Pharmacol.* **57,** 1–15.

Kagawara Y. and Racker E. (1971) Partial resolution of the enzymes catalyzing oxidative phosphorylation. *J. Biol. Chem.* **246,** 5477–5487.

Kanner B. I. (1978a) Solubilisation and reconstitution of the γ-aminobutyric acid transporter from rat brain. *FEBS Lett.* **89,** 47–50.

Kanner B. I. (1978b) Active transport of γ-aminobutyric acid by membrane vesicles isolated from rat brain. *Biochemistry* **17,** 1207–1211.

Kohn M. C., Menten L. E., and Garfinkel D. (1979) A convenient computer program for fitting enzymatic rate laws to steady state data. *Comput. Biomed. Res.* **12,** 461–469.

Lahdesmaki P. and Oja S. S. (1973) Mechanism of taurine transport at brain cell membranes. *J. Neurochem.* **20,** 1411–1417.

Lajtha A. (1983) *Handbook of Neurochemistry,* Vol. 5, 2nd Ed., Plenum, New York.

Lake N. and Voaden M. J. (1976) Exchange versus net uptake of exogenously-applied γ-aminobutyric acid in retina. *J. Neurochem.* **27,** 1571–1573.

Lakshmanan J. and Padmanaban G. (1974) Effect of α-oxalyl-L-α, β-diaminopropionic acid on glutamate uptake by synaptosomes. *Nature* **249,** 469–471.

Larsson O. M., Johnston G. A. R., and Schousboe A. (1983) Differences in uptake kinetics of cis-3-aminocyclohexane carboxylic acid into neurons and astrocytes in primary cultures. *Brain Res.* **260,** 279–285.

Larsson O. M., Thorbek P., Krogsgaard-Larsen P., and Schousboe A. (1981) Effects of homo-β-proline and other heterocyclic GABA analogues on GABA uptake in neurons and astroglial cells and GABA receptor binding. *J. Neurochem.* **37,** 1509–1516.

Lasher R. S. (1974) The uptake of [^{3}H] GABA and differentiation of stellate neurons in cultures of dissociated postnatal rat cerebellum. *Brain Res.* **69,** 235–254.

Latorre R., Coronado R., and Vergara C. (1984) K$^+$ channels gated by voltage and ions. *Ann. Rev. Physiol.* **46,** 485–495.

Levi G., Banay-Schwartz M., and Raiteri M. (1978) Uptake, Exchange, and Release of GABA in Isolated Nerve Endings, in *Amino Acids as Chemical Transmitters* (Fonnum F., ed.) Plenum, New York, pp 327–350.

Levi G., Bertollini A., Chen J., and Raiteri M. (1974) Regional differences in the synaptosomal uptake of ^{3}H-γ-aminobutyric acid and ^{14}C-glutamate and possible role of exchange processes. *J. Pharmacol. Exp. Therap.* **188,** 429–438.

Levi G., Coletti A., Poce V., and Raiteri M. (1976) Decrease of uptake and exchange of neurotransmitter amino acids after depletion of their synaptosomal pools. *Brain Res.* **103,** 103–116.

Levi G. and Raiteri M. (1973) GABA and glutamate uptake by subcellular fractions enriched in synaptosomes: Critical evaluation of some methodological aspects. Brain Res. **57,** 165–185.

Levi G. and Raiteri M. (1974) Exchange of neurotransmitter amino acids at nerve endings can stimulate high-affinity uptake. *Nature* (Lond.) **250,** 735–737.

Levi G. and Raiteri M. (1976) Synaptosomal transport processes. *Int. Rev. Neurobiol.* **19,** 51–74.

Levy W. B., Haycock J. W., and Cotman C. W. (1976) Stimulation-dependent depression of readily releasable neurotransmitter pools in brain. *Brain Res.* **115,** 243–256.

Levy W. B., Redburn D. A., and Cotman C. W. (1973) Stimulus-coupled secretion of γ-aminobutyric acid from rat brain synaptosomes. *Science* **181,** 676–678.

Lilly J. C. (1961) Injury and Excitation by Electric Currents. A. The Balanced Pulse-Pair Waveform, in *Electrical Stimulation of the Brain* (Sheer D. E., ed.) University of Texas Press, Austin, pp 60–64.

Martin D. L. (1976) Carrier-Mediated Transport and Removal of GABA from Synaptic Regions, in *GABA in Nervous System Function* (Roberts E., Chase T. N., and Tower D. B., eds.) Raven Press, New York, pp. 347–386.

Martin D. L. and Smith A. A. (1972) Ions and the transport of gamma-aminobutyric acid by synaptosomes. *J. Neurochem.* **19,** 841–855.

McGeer P., Eccles J. C., and McGeer E. G. (1978) *Molecular Neurobiology of the Mammalian Brain,* Plenum, New York.

McIlwain H. (1975) *Practical Neurochemistry,* 2nd Ed., Churchill Livingstone, Edinburgh.

McIlwain H. and Rodnight R. (1962) *Practical Neurochemistry,* Churchill, London.

Minchin M. C. W. (1975) Factors influencing the efflux of [^{3}H] gamma-aminobutyric acid from satellite glial cells in rat sensory ganglia. *J. Neurochem.* **24,** 571–577.

Minchin M. C. and Iversen L. L. (1974) Release of [^{3}H] gamma-aminobutyric acid from glial cells in rat dorsal root ganglia. *J. Neurochem.* **23,** 533–540.

Morgan I. G. (1976) Synaptosomes and cell separation. *Neuroscience* **1,** 159–165.

Morton I. K. M., Stagg C. F., and Webster R. A. (1977) Perfusion of the central canal and subarachnoid space of the cat and rabbit spinal cord in vivo. *Neuropharmacol.* **16,** 1–6.

Neal M. J. and Bowery N. G. (1979) Differential effects of veratridine and potassium depolarization on neuronal and glial GABA release. *Brain Res.* **167,** 337–343.

Neal M. J. and Starr M. S. (1973) Effects of inhibitors of γ-aminobutyrate aminotransferase on the accumulation of ^{3}H-γ-aminobutyric acid by the retina. *Brit. J. Pharmacol.* **47**, 543–555.

Oja S. S. and Korpi E. R. (1983) Amino Acid Transport, in *Handbook of Neurochemistry*, 2nd Ed., Vol. V (Lajtha A., ed) Plenum, New York, pp 311–337.

Orrego F. (1979) Criteria for the identification of central neurotransmitters, and their application to studies with some nerve tissue preparations in vitro. *Neuroscience* **4**, 1037–1057.

Orrego F., Jankelevich J., Ceruti L., and Ferrara E. (1974) Differential effects of electrical stimulation on release of ^{3}H-noradrenaline and ^{14}C-α-aminoisobutyrate from brain slices. *Nature* (Lond.) **251**, 55–57.

Orrego F. and Miranda R. (1976) Electrically induced release of (^{3}H)-GABA from neocortical thin slices. Effects of stimulus waveform and of amino-oxyacetic acid. *J. Neurochem.* **26**, 1033–1038.

Orrego F., Miranda R., and Saldate C. (1976) Electrically induced release of labeled taurine, α- and β-alanine, glycine, glutamate, and other amino acids from rat neocortical slices in vitro. *Neuroscience* **1**, 325–332.

Pappius H. M. and Elliott K. A. C. (1956) Water distribution in incubated slices of brain and other tissues. *Can. J. Biochem. Physiol.* **34**, 1007–1022.

Pastuszko A., Wilson D. F. and Erecinska M. (1981) Net uptake of γ-aminobutyric acid by a high-affinity system of rat brain synaptosomes. *Proc. Natl. Acad. Sci. USA* **78**, 1242–1244.

Patterson M. S. and Greene R. C. (1965) Measurement of low energy beta-emitters in aqueous solution by liquid scintillation counting of emulsions. *Anal. Chem.* **37**, 854–857.

Pearce B. R., Currie D. N., Beale R., and Dutton G. R. (1981) Potassium-stimulated, calcium-dependent release of [^{3}H] GABA from neuron- and glia-enriched cultures of cells dissociated from rat cerebellum. *Brain Res.* **206**, 485–489.

Pearce B. R., Currie D. N., Dutton G. R., Hussey R. E. G., Beale R., and Pigott R. (1981) A simple perfusion chamber for studying neurotransmitter release from cells maintained in monolayer culture. *J. Neurosci. Methods* **3**, 255–259.

Raiteri M., Frederico R., Coletti A., and Levi G. (1975) Release and exchange studies relating to the synaptosomal uptake of GABA. *J. Neurochem.* **24**, 1243–1250.

Ramaharobandro N., Borg J., Mandel P. and Mark J. (1982) Glutamine and glutamate transport in cultured neuronal and glial cells. *Brain Res.* **244**, 113–121.

Redburn D. A., Biela J., Shelton D. L. and Cotman C. W. (1975) Stimulus secretion coupling in vitro: A rapid perfusion apparatus for monitoring efflux of transmitter substances from tissue samples. *Anal. Biochem.* **67**, 268–278.

Reubi J. and Cuenod M. (1976) Release of exogenous glycine in the pigeon optic tectum during stimulation of a midbrain nucleus. *Brain Res.* **112**, 347–361.

Roberts P. J. (1974) Amino acid release from isolated rat dorsal root ganglia. *Brain Res.* **74,** 327–332.

Roberts P. J. (1976) Gamma-aminobutyric acid homoexchange in sensory ganglia. *Brain Res.* **113,** 206–209.

Roskoski R. (1978) Net uptake of L-glutamate and GABA by high-affinity synaptosomal transport systems. *J. Neurochem.* **31,** 493–498.

Rubin R. P. (1970) The role of calcium in the release of neurotransmitter substances and hormones. *Pharmacol. Rev.* **22,** 389–498.

Ryan L. D. and Roskoski R. (1975) Selective release of newly synthesized and newly captured GABA from synaptosomes by potassium depolarization. *Nature* (Lond.) **258,** 254–256.

Ryan L. D. and Roskoski R. (1977) Net uptake of GABA by a high-affinity synaptosomal system. *J. Pharmacol. Exp. Therap.* **200,** 285–291.

Salceda R. and Pasantes-Morales H. (1975) Calcium coupled release of (^{35}S) taurine from retina. *Brain Res.* **96,** 206–211.

Schlaepfer W. W. (1977) Structural alterations of peripheral nerve induced by the calcium ionophore A 23187. *Brain Res.* **136,** 1–9.

Schon F. and Kelly J. S. (1974) The characterization of [^{3}H]-GABA uptake into the satellite glial cells of rat sensory ganglia. *Brain Res.* **66,** 289–300.

Schousboe A. (1981) Transport and metabolism of glutamate and GABA in neurons and glial cells. *Int. Rev. Neurobiol.* **22,** 1–45.

Schousboe A. (1982) Metabolism and Function of Neurotransmitters, in *Neuroscience Approached Through Cell Culture,* Vol. I (Pfeiffer S. E., ed.) CRC Press, Boca Raton, pp. 107–141.

Schousboe A., Larsson O. M., Drejer P., Krogsgaard-Larsen P., and Hertz L. (1983) Uptake and Release Processes for Glutamine, Glutamate, and GABA in Cultured Neurons and Astrocytes, in *Glutamine, Glutamate, and GABA in the Central Nervous System* (Hertz L., Kvamme E., McGeer E., and Schousboe A., eds.) Alan R. Liss, New York, pp. 297–315.

Schousboe A., Larsson O. M., Hertz L., and Krogsgaard-Larsen P. (1981) Heterocyclic GABA analogs as new selective inhibitors of astroglial GABA transport. *Drug Dev. Res.* **1,** 115–127.

Schousboe A., Thorbek P., Hertz L., and Krogsgaard-Larsen P. (1979) Effects of GABA analogs of restricted conformation on GABA transport in astrocytes and brain cortex slices and on GABA receptor binding. *J. Neurochem.* **33,** 181–189.

Schrier B. K. and Thompson E. J. (1974) On the role of glial cells in the mammalian nervous system. Uptake, excretion, and metabolism of putative neurotransmitters by cultured glial tumor cells. *J. Biol. Chem.* **249,** 1769–1780.

Schwartzkroin P. A. (1981) To Slice or Not To Slice, in *Electrophysiology of Isolated Mammalian CNS Preparations* (Kerkut G. A. and Wheal H. V., eds.) Academic Press, London, pp 15–50.

Scott-Young W. (1985) In Vivo Autoradiographic Localization of Amino Acid Receptors and Uptake Sites, in *Neuromethods: Amino Acids*

(Boulton A. A., Baker G. B., and Wood J. D., eds.) Humana, Clifton, New Jersey (in press).

Sellstrom A. and Hamberger A. (1977) Potassium-stimulated γ-aminobutyric acid release from neurons and glia. *Brain Res.* **119,** 189–198.

Sellstrom A., Venema R. and Henn F. (1976) Functional assessment of GABA uptake or exchange by synaptosomal fractions. *Nature* (Lond.) **264,** 652–653.

Sershen H. and Lajtha A. (1974) The distribution of amino acids, Na$^+$ and K$^+$ from surface to center in incubated slices of mouse brain. *J. Neurochem.* **22,** 977–985.

Shank R. P. and Campbell G. L. (1984) Amino acid uptake, content, and metabolism by neuronal and glial enriched cellular fractions from mouse cerebellum. *J. Neurosci.* **4,** 58–69.

Shank R. P. and Baxter C. F. (1975) Uptake and metabolism of glutamate by isolated toad brains containing different levels of endogenous amino acids. *J. Neurochem.* **24,** 641–646.

Simon J. R., Martin D. L., and Kroll M. (1974) Sodium-dependent efflux and exchange of GABA in synaptosomes. *J. Neurochem.* **23,** 981–991.

Szerb J. C. (1982) Effect of nipecotic acid, a γ-aminobutyric acid transport inhibitor, on the turnover and release of γ-aminobutyric acid in rat cortical slices. *J. Neurochem.* **39,** 850–858.

Szerb J. C. (1983) Mechanisms of GABA Release, in *Glutamine, Glutamate, and GABA in the Central Nervous System* (Hertz L., Kvamme E., McGeer E., and Schousboe A.,eds.) Alan R. Liss, New York, pp. 457–472.

Tang C. M., Cohen M. W., and Orkand R. K. (1979) Sodium channels in axons and glial cells of the optic nerve of *Necturus maculosa. J. Gen. Physiol.* **74,** 629–642.

Tapia R. (1983) γ-Aminobutyric acid. Metabolism and Biochemistry of Synaptic Transmission, in *Handbook of Neurochemistry* Vol. 3 (Lajtha L., ed.) Plenum, New York, pp 423–466.

Ulbricht W. (1969) The effect of veratridine on excitable membranes of nerve and muscle. *Ergeb. Physiol. Biol. Chem. Exp. Pharmakol.* **61,** 18–71.

Valdes F. and Orrego F. (1978) Electrically induced calcium-dependent release of endogenous GABA from rat brain cortex slices. *Brain Res.* **141,** 357–363.

Vargas O., de Lorenzo M. D. C., and Orrego F. (1977a) Effect of elevated extracellular potassium on the release of labeled noradrenaline, glutamate, glycine, β-alanine, and other amino acids from rat brain cortex slices. *Neuroscience* **2,** 383–390.

Vargas O., de Lorenzo M. D. C., Saldate M. C., and Orrego F. (1977b) Potassium-induced release of [^{3}H]-GABA and of [^{3}H]-noradrenaline from normal and reserpinized rat brain cortex slices. Differences in calcium-dependency, and in sensitivity to potassium ions. *J. Neurochem.* **29,** 165–170.

Vargas O., Miranda R., and Orrego F. (1976) Effects of sodium-deficient media and of a calcium ionophore (A-23187) on the release of [^{3}H]-noradrenaline, [^{14}C]-α-aminoisobutyrate, and [^{3}H]-γ-aminobutyrate from superfused slices of rat neocortex. *Neuroscience* **1**, 137–145.

Vargas O. and Orrego F. (1976) Elevated extracellular potassium as a stimulus for releasing [^{3}H]-norepinephrine and [^{14}C]-α-aminoisobutyrate from neocortical slices. Specificity and calcium-dependency of the process. *J. Neurochem.* **26**, 31–34.

Varon S., Weinstein H., Baxter C. F., and Roberts E. (1975) Uptake and metabolism of exogenous γ-aminobutyric acid by subcellular particles in a sodium-containing medium. *Biochem. Pharmacol.* **14**, 1755–1764.

White R. D. and Neal M. J. (1976) The uptake of L-glutamate by the retina. *Brain Res.* **111**, 79–93.

Winegard S. and Shanes A. M. (1962) Calcium flux and contractility in guinea pig atria. *J. Gen. Physiol.* **45**, 371–394.

Yu A. C. H. and Hertz L. (1982) Uptake of glutamate, GABA, and glutamine into a predominantly GABA-ergic and a predominantly glutamergic nerve cell population in culture. *J. Neurosci. Res.* **7**, 23–35.

Index

A-23187, 250

AAT (*see* Aspartate amino transferase)

Acetate, 74, 204, 205, 210, 213, 215

Acetone, 3, 33

Acetonitrile, 33, 41

Acetylcholine, 171, 202

Acetyl CoA, 222

N-Acetyl aspartyl-L-glutamate (AAG), 144, 146

γ-Acetylenic-GABA, 219

Acid phosphatase, 155

Acylation, 35–42, 53–64

β-Adrenergic blockers, 145

AIB (*see* α-Aminoisobutyric acid)

Ala (*see* Alanine)

Alanine (Ala), 10–14, 31, 100–104, 107, 110, 111

β-Alanine (β-Ala), 102, 106–108, 246

Alanine aminotransferase, 215, 218

Alkoxycarbonylation, 38–40

Alkyl chloroformate, 39, 40

Allo-isoleucine, 5, 19

Amino acids in
blood, 7, 8, 10, 33, 34, 44, 66
central nervous system, 6, 7, 32, 33, 66–68, 83, 84, 91, 92, 103, 104, 107–111, 134, 135, 166, 168, 170, 186–195, 202, 211, 224, 225, 258–260
cerebrospinal fluid, 8, 9, 11, 33, 34, 111, 112, 259, 260
urine, 9–13, 33, 34, 44, 66

visual system, 106–108, 167, 169, 261, 262

α-Aminoadipic acid, 102, 144

α-Amino-*n*-butyric acid, 10, 11, 13, 102, 110

β-Aminobutyric acid, 13, 102, 104

γ-Aminobutyric acid (GABA), 3, 6, 14, 29, 31, 32, 52–58, 64–67, 69–74, 84, 85, 87, 88, 91–93, 101, 102, 104, 106, 107, 110, 111, 121, 125, 129, 133, 135, 138–142, 144–147, 156, 171, 180, 183, 185–190, 195, 201–205, 207–210, 212–223, 225, 226, 239, 241–246, 248, 251, 252, 254, 256, 258, 260–263

cis-3-Aminocyclohexane carboxylic acid, 246

2-Amino-3-guanidino propionic acid, 3

α-Amino-3-hydroxy-5-methylisoxazole-4-propionic acid (AMPA), 194

α-Aminoisobutyric acid, 102, 243, 252, 253

6-Aminonicotinamide, 212

Amino-oxyacetic acid (AOAA), 215–217, 220, 226, 242, 254

2-Amino-4-phosphonobutyric acid (APB), 135, 191

2-Amino-7-phosphonoheptanoic acid (APH), 128, 146, 214

Aminopropane sulfonic acid, 124

Ammonia, 9, 14

Ammonium acetate, 212

AMPA (*see* α-Amino-3-hydroxy-5-

273